T0134202

The Apollo Chronicles

The Apollo Chronicles

Engineering America's First Moon Missions

Brandon R. Brown

OXFORD
UNIVERSITY PRESS

Oxford University Press is a department of the University of Oxford. It furthers the University's objective of excellence in research, scholarship, and education by publishing worldwide. Oxford is a registered trade mark of Oxford University Press in the UK and certain other countries.

Published in the United States of America by Oxford University Press
198 Madison Avenue, New York, NY 10016, United States of America.

CIP data is on file at the Library of Congress
ISBN 978-0-19-068134-0

For my father and my mother.

For the Apollo engineers and their families.

CONTENTS

Preface ix

Acknowledgments xiii

1. 1945—Origins 1

2. 1957—Paths, Power, and Panic 12

3. 1960—Silent Movies and Old-World Evenings 24

4. 1961—A Toddler's Marathon 42

5. The Moon 60

6. 1962—Punch Cards and a Key to the Trunk 64

7. 1964—Of Doubts and Bugs 80

8. 1965—Saturn Breathes 99

9. 1966—Of Software and Star Balls 115

10. 1967—From Madness to Miracle 133

11. 1968—Of Timeless Views and New Perspectives 147

12. 1969—Alarms and Lightning 170

13. 1972—From Rovers to Regrets 197

14. 1981—Farther Along 215

15. Today—Mementos and Returns 222

16. How We Did It 230

Notes 241
Bibliography 255
Index 259

PREFACE

On a winter's night three years ago, I looked up at the full Moon, with its dark and light patches, and realized all at once that I knew very little about my father's early career. Almost nothing. I'd met many of his fellow engineers, heard stories of their quirks and their humor, but I had no idea what living and working the Moon missions was like in the 1960s. As of this writing, my father has just turned eighty-five. He has some very sharp memories of certain segments and certain projects, while other parts are understandably hazy now. After I started peppering him with questions, he brought out a box of keepsakes and memorabilia from his career days, talking through a review of each piece. He also shared a few phone numbers with me, of surviving colleagues. And when I showed him a list of names from NASA's archive of oral histories, he gamely circled those he knew would have interesting and important stories to tell.

I started reading everything I could find about Apollo—America's program to land people on the Moon and return them alive to our planet's welcoming surface—and the steps that came before it. Many wonderful narratives focus on the astronauts and the blow-by-blow missions in space. But I mostly wanted to understand the project's earthly trenches, with engineers sweating details, deadlines, and decisions of a sort no person had faced before. In the eleven years spanning the formation of NASA in 1958 to the first Moon landing in 1969, the engineers cleared or circumvented every hurdle. If they did not solve every problem, they solved most and cleverly soothed the rest. Another set of books, fewer in number, detail the engineering work, including the piles of acronyms, the shifting, expanding organizational charts, the precise titles and versions of the various systems of each rocket, spacecraft, testing platform, and so on. The combination of NASA's strange internal parlance and Apollo's monumental technical complexity raises a significant hurdle for many readers who might otherwise want to understand the work, the process, and the experience. What was it really like, day to day and month by hectic month?

The Apollo engineer Peter Armitage once told an interviewer, "The real story is in the people and why they behaved the way they behaved. Nobody's ever written a book like that yet—the real people, the mistakes they made." With humility, I have tried to write a book that could leave the remaining engineers nodding in approval. But

I also have written this book for anyone my age or younger. I was born in early 1969, months before the first Moon landing, and by the time my generation became aware of the world, Apollo was old news.

This book is not a memoir. As the least interesting element involved, your author will now recede. Neither is it my father's story, though a few of his anecdotes will improve the pages that follow. And while I now appreciate just how brave our astronauts were, to sit on towers of explosive fuel and venture into a deadly realm, this book is more concerned with the astronauts' protectors. In it, I focus on the Earth-bound: the welders of space-worthy seams, the designers of heat shields, the stitchers of spacesuits, and those who computed razor-thin trajectories through space, with disaster awaiting any deviation or missed step. As you read, you'll sit with the mystified engineers who, barely out of college, watched warning lights blink on at the worst possible times. You'll hunker with the rocket engineers in the shaking walls of a block house during tests of the world's most powerful rocket engines. You'll crawl with draftsmen over sprawling, improvised tables, drawing around the clock to complete hundreds of schematics. And you'll sit with young, farm-raised Americans learning rocketry from imported German experts.

The stories, even the ones we haven't yet lost, are nearly endless. Any attempt to comprehensively honor the four hundred thousand minds that pioneered the missions would be as impossible to assemble and write as it would then be to read. You will meet a number of the engineers but avoid long parades of names. This book will explore the Apollo years using a handful of main characters like blood cells moving through the various limbs of NASA. I want to urge the reader against the impression that these individuals were super-human. Two were most definitely visionaries, but I've selected others because they were involved in key milestones, worked in multiple parts of NASA, and, crucially, have retained detailed memories of the Apollo era.

The first act of a dramatic arc to the Moon began in October of 1957. When the Soviet Union put a metallic "artificial moon" in orbit around the Earth, Americans quite nearly lost their minds. What did it mean that our enemies, at unprecedented speeds and heights, could methodically paint these bands—these *orbits*—again and again over our skies? Could the Sputnik see us? Was it a hostile device? Could it drop an atomic bomb? As the political historian Walter McDougall wrote, "For the first time since [the War of 1812] the American homeland lay under direct foreign threat. . . ." Newspapers compared flat-footed America to "some fat Roman lolling in the baths" before a barbarian invasion.[1]

Never mind catching up, the nation thought. How could we get into space at all? What did the country have in its engineering cupboard? If nameless Soviet scientists were plotting even larger and more menacing missions, who would lead the way for the humiliated United States?

Before experiencing 1957, you will meet some of our cast in earlier times. In most cases, they had little inkling that history would sweep them up and push them onto a brightly lit technical stage. The world would soon watch them working in a theater of bewildering scale and scope. When the curtain rushed open, they would stand without script or rehearsal, looking to one another for whatever relevant experience they could muster. And to a surprising extent, humanity's peaceful path to the Moon relied on ideas born in war.

July 2018

ACKNOWLEDGMENTS

I first want to thank the engineers and scientists I have interviewed. In sharing their recollections and expertise, the following people breathed great life info this project: Bob Austin, Hal Beck, Eugene Benton, Aldo Bordano, Robert Brown, Marlowe Cassetti, Nesbitt Cumings, Caleb Fassett, Gerry Griffin, Mack Henderson, Frank Hughes, John Kastanakis, Ed Kowalchuk, Arnolia McDowell, Elric McHenry, Jack Miller, Thomas Moser, Debra Needham, Lee Norbraten, Catherine Osgood, Thomas Parnell, Henry Pohl, Wesley Ratcliff, William Sneed, Ken Young, Renee Weber, Cynthia Wells, Don Woodruff, and Len Worlund.

I want to acknowledge Hal Beck, Lee Norbraten, Thomas Parnell, and Henry Pohl for sharing some of their unpublished writings about their careers.

For providing further background and for their interview time, I thank Ann Faget, Carol Faget, Guy Faget, Chip Lord, and Karl Pohl.

For historical and archival assistance, I gratefully acknowledge Brian Odom, Historian at the Marshall Spaceflight Center in Huntsville, Alabama; Pat Ammons and Carolyn Lawson of the United States Space and Rocket Center in Huntsville, Alabama; and Steve Garber and Bill Barry of the NASA History Office in Washington, D.C.

In terms of source material, I acknowledge all those who have compiled and documented so very much already, and I apologize to those I have yet to discover. I leaned heavily on NASA's Oral Histories project, incorporating dozens of voices from those digital archives. The book *Apollo: The Behind-the-Scenes History of the Most Triumphant Years of America's Space Program*, by Charles Murray and Catherine Bly Cox, was a wonderful companion on my learning curve. Finally, a lesser known tome, *Dr. Space: The Life of Wernher von Braun* by the late Bob Ward, provided an intimate portrait of the rocket pioneer and his center in Huntsville, Alabama.

I want to gratefully acknowledge the patient and eagle-eyed readers of various drafts: Dana Smith, Sue Brown, Arden Hendrie, and Dean Rader, as well as, from the Apollo era, Robert Brown, Gerry Griffin, and Frank Hughes. All of these people improved the book, and any remaining errors are mine alone.

I appreciate Jeremy Lewis, my editor at Oxford University Press, very much for his trust, support, and astute suggestions. I also want to acknowledge the expert and timely assistance provided by Book of Earl Indexing. As always, I thank my agent, Jennifer Lyons, for her guidance, encouragement, and brainstorming.

Some research costs were covered by the University of San Francisco's Faculty Development Fund. I also appreciate the supportive community of my department at the University.

Finally, special thanks go to my wife, Dana, for her daily patience and reflection as I nattered about this project. Olive the whippet also endured it, with many a heavy sigh.

The Apollo Chronicles

1

1945—ORIGINS

After surrendering to American forces in May of 1945, the lead scientist of Nazi Germany's rocketry program prepared to embark for the United States. Wernher von Braun cooperated with his captors. He alternately charmed and shocked them with his cheerful good will and confident self-importance. (In truth, he watched his plan coming together. As he'd confided to a few fellow Germans late in the war, he had hoped to end up in the United States.) He asked if he could bring hundreds of his engineers and workers with him to America. He amazed his interrogators as he rattled off the strengths, skills, and weaknesses of *each* man. Washington eventually approved 118, who came to America in two waves.

In September, von Braun and a handful of his closest German colleagues deplaned at the Newcastle Army Base in Wilmington, Delaware. At age thirty-three, his first American steps were painful ones. After surviving a near-fatal car wreck in Germany, the rocket pioneer suffered from a poorly set broken arm and a nasty case of hepatitis. His new military keepers let von Braun recover his strength for a couple of weeks but then covertly booked him a civilian train ticket. When his accent attracted the questions of a fellow traveler, von Braun said he was Swiss and concocted business interests in the states.

From his window, von Braun saw the vast expanse of a nation rumble past for days. Completely unlike the craters and ash piles of Europe, this land was free of war's scars. The trees thinned as the train clacked onward and eventually, in the middle of Texas, the grass gave up as well. He arrived in El Paso, a town perched between two stands of dry mountains. Fort Bliss became the new home for von Braun and his German colleagues. The enlisted locals deduced the basics about their new prisoners-turned-guests: These were the infamous enemy scientists, plucked from Nazi Germany and set in the cage of a rocket laboratory. They jokingly called von Braun "the Dutchman."

The initially secret Project Overcast hauled ten thousand tons of military and industrial equipment, about seventy-five rockets (many of them in piles of parts), and six hundred former enemies from Germany to the United States, largely whitewashing their political affiliations and wartime work. In von Braun's case, the cleansing covered, for many years to come, his advanced rank within the Nazi war machine. He later claimed a tin ear for politics, calling himself "downright naïve" in his early political views, blind to the meaning of the changes around him. He said he was simply "too wrapped up in rockets." When the Nazis took control of Germany, he was barely twenty-one. He joined the Nazi party and eventually became an officer in the Nazi SS (*Schutzstaffel*, or "protection squad"). Heinrich Himmler's brutal wing ran the concentration camps and, for von Braun, they eventually supplied slave labor—thousands of prisoners of war and other enemies of the state toiling to build von Braun's massive rocket weapons.[1]

According to the story he relayed in America, his Nazi masters had tilted his career away from his visions of space travel in favor of earthly and more nefarious trajectories. He said he always would have preferred to perch spaceships, rather than bombs, atop his rockets. He had even lapsed late in World War II, dangerously wishing aloud for the war to end so he could return to his real passions. He possessed an almost deadly naïveté, scribbling sarcastic comments like "Final victory? Well, well!" in his notes, mocking the late war slogans of Josef Goebbels. The Gestapo arrested von Braun in early 1944 and charged him with scheming to subvert the Nazi weapons program. If not for his skills, plus some friends carefully extracting him from the Gestapo's bureaucracy, the von Braun story could have ended there. Under increased scrutiny, he followed orders and survived the war, with rocket schematics tucked under his arm.[2]

Once Project Overcast became public in America, it took on the more benign name "Project Paperclip," and newspapers even printed quiet notice of von Braun's arrival. He was just one of the "certain outstanding German scientists and technicians . . . deemed vital to our national security" who would have a "temporary stay." For von Braun that would span the rest of his life.

At Fort Bliss, von Braun and his German colleagues expected tension with American soldiers, or perhaps retribution for years of war. But he found the opposite. The soldiers invited him to join their card games. "In America," he later wrote, "you don't seem to carry grudges, as do many Europeans who have been enemies." He encouraged his fellow Germans to learn English as quickly and thoroughly as possible. They watched whatever movies they could and listened to the radio. Not long after they arrived, some Germans took to wearing cowboy hats and boots.[3]

But before he could build new rockets, the Americans wanted to see his most famous work up close: He and his fellow Germans trudged into New Mexico's nearby white sands, with loads of their rocket parts. He was to revive his chief accomplishment for Adolf Hitler, a weapon of incredible size and speed.

From von Braun's perspective, he could help the United States develop long-range missiles, yes, but with the victors' resources he might return to his visions of space travel. With enough time and money, he was certain he could build a rocket powerful enough to refuse Earth's gravity and glide calmly through the cosmos. So he made a new home willingly and joined his second army within one year. "It all made sense," he said later. "The V-2 was something we had and you didn't have."[4] (See Figure 1.1.)

The Vergeltungswaffe-2 was the second of Nazi Germany's "vengeance weapons." The desperate Reich had hoped the V-2 would reverse the late course of the war.

When preparing for launch, it stood forty-five feet tall, tethered like some sort of laboratory monster. Oozing vapors, it was part science fiction and part Gothic horror. If the rocket would launch with flame, why did tendrils of super-cold mists hang from its fuel lines?

As with most combustible materials, the liquid fuel in the V-2 required oxygen to light and burn, but the air around us only offers so much. And at the heights von Braun craved, oxygen was incredibly scarce. Von Braun had his rockets carry their

FIGURE 1.1 German and American workers prepare a V-2 rocket for a test at the White Sands Proving Grounds in New Mexico, circa 1946. (Photograph courtesy the U.S. Space and Rocket Center.)

own oxygen supply: five tons compressed and liquefied at a chilly −300° Fahrenheit. The use of liquid oxygen predated von Braun. As a student, he had absorbed the speculative idea while reading *The Rocket into Interplanetary Space* by his eventual mentor, Hermann Oberth. In the early 1920s, Oberth had exchanged letters with the American physicist and rocket pioneer Robert Goddard. And, largely unknown to those tinkering dreamers, the Russian schoolteacher and visionary Konstantin Tsiolkovsky had already proposed using liquefied oxygen many years before.[i] As with so many technologies, when civilization was ready to make a leap, an idea bubbled forth in many places and from many minds.[5]

Rendering all previous work quaint by comparison, von Braun's engines pumped oxygen to meet ethanol in a fiery brew. "The whole sky seemed to vibrate," said one team member of their first successful V-2 test. "This kind of unearthly roaring was something human ears had never heard." They understood they had sparked a new era. As their rockets sped from our thin atmosphere and first grazed what lay beyond, they started humanity's journey into space.[6]

Germany's rocket weapons, launched by the hundreds against Belgium and southern England, alarmed the Allies, but the sonic boom of the rockets proved worse than their bite. Frequently inaccurate, they had little effect on the course of the war, but they indirectly shifted the fate of an American family and the nation itself.

Aiming to defang Germany's rocket threat, the Allies hatched an ill-conceived marriage of self-guided rockets with Japan's *kamikaze* attacks. *Operation Aphrodite* was a failed U.S. attempt to have small crews take off in explosive-packed bombers, turn piloting over to radio remote control, and parachute to safety before the drone-like vehicles struck German rocket facilities. An experienced American pilot named Joseph Kennedy, Jr., volunteered for the risky missions. As he prepared to leap from one such bomber, his cargo detonated prematurely. The heir apparent to the Kennedy family's political ambitions died in a fireball over the English Channel.

In the summer of 1945, with von Braun starting a new life in America, a younger Kennedy walked through the ashen ruins of Germany. At the age of twenty-eight, John had just started his first civilian post. After surviving combat in the Pacific, he flexed his father's connections. Hearst newspapers placed him in a prime role, documenting the war's aftermath in Europe. But his private journal reveals less about the news of 1945 and more about a keen soul discerning the world's shifting politics. John F. Kennedy looked at smoldering central Europe, with its new partitions, and acknowledged a new, colder war. "I do not agree with those people who advocate war now with the Russians on the argument of 'Eventually, why not now?'," he wrote. He

---

[i] Tsiolkovsky's inspiration, in turn, sprang from the space-faring novels of the French writer Jules Verne.

predicted that a new technological superweapon could put the next major war on in-definite hold. These journal entries preceded the atomic explosions in Hiroshima and Nagasaki by mere weeks. "Thus Science," he mused, "which has contributed so much to the horrors of war, will still be the means of bringing it to an end."[7]

Also in the summer of 1945, Maxime "Max" Faget toiled in another theater of war and in a very different environment. He glided through the dark depths of the Pacific Ocean. As a junior officer on the submarine USS Guavina, he returned, he presumed, to another terrifying battle with the Japanese navy. Short, lively, and shrewd, he had chosen submarine duty over other military service for one primary reason: He had never met a veteran of submarine battle with tragic injuries. It was all or nothing—a simple death or unscathed survival (see Figure 1.2).

Given that his father, the doctor Guy Henry Faget, researched tropical diseases, Max had been born in Belize and raised in British Honduras. When Dr. Faget sub-sequently decided to focus on a new treatment for leprosy, he returned the family to their southern roots. In Carville, Louisiana, he took over one of the nation's largest leper hospitals. Son Max earned his mechanical engineering degree from Louisiana State University before joining the war effort.

FIGURE 1.2 Maxime "Max" Faget in 1944. (Photograph courtesy the Faget family.)

By August of 1945, the Guavina had already seen some of the most intense action of any U.S. submarine. After recording a number of high-profile hits on a Japanese convoy in February of 1945, the Guavina barely survived an intense seven hours of Japanese retaliation, including nearly one hundred depth charges. In relatively shallow water, the Guavina could only lie silently on the sandy bottom, hoping none of those charges would tear it open. Each underwater explosion was, according to Faget, "like a sledgehammer slamming the hull next to you."[8]

Soon thereafter, the Navy alerted the Guavina of a ditching American bomber. As soon as the sub surfaced, Faget dove into the water to save a number of stunned airmen from drowning, earning him Navy commendations for bravery.

The submarine life taught Max Faget about survival in a hostile environment, a place humans were never meant to be. A machine could hold enough air around a crew to keep them alive for a surprisingly long time, even with violence swirling outside. And the submarine required a technological awareness. It had to navigate murky or sometimes pitch black realms. These long, tense hours planted seeds in Faget's unique engineering imagination, one that would someday design ships that transformed impossible goals to practical simplicity.

This life at sea also provided moments of wonder. He later told his children about a night, on the surface, when he was keeping watch. It was a moonless night, with the water lit only by stars, and he grew nervous when he saw a bright spot growing on the Eastern horizon. If this was another Japanese squadron, the Guavina would need to dive. But no, he could see a pinpoint rising now and its orange glow spread across the calm ocean. Jupiter, king of the planets, marked a path through space and cared not one bit about little Earth or humanity's sad wars.[9]

What Faget would *not* carry with him from his military days was any love of authority or concern for hierarchy. In his engineering work to come, his colleagues quickly came to recognize not just his quirky brilliance but also a requisite long leash. He would do things his own way, often sinking into what they called a "trance" for days on end until a new design fully revealed itself.[10]

Von Braun and Faget, two veterans and former enemies, adjusted to life after war, and each hungered for new peacetime challenges. They would become central cogs in America's space-faring efforts. Other minds, by the hundreds of thousands, would also join this push to leave Earth and its atmosphere behind. In 1945, most of them were just children or teenagers.

At war's end, a twelve-year-old named Henry Pohl worked on his family's farm in rural Texas, and his only aim was to someday have his own land and try his hand at agriculture. During World War II, his father made a luxury purchase, a three-dollar radio, and listened to war reports during midday lunch breaks. Neighbor families would walk over on Saturday nights to listen to the radio with the Pohls. Henry burned out their first battery (advertised to last one year) in just a few weeks because he couldn't get enough of it. Devices captivated the young Pohl, and he still recalls the precise steps and tools his family used to dig a well, cut beams, or level a house.

The Pohls lived in what the locals called "the summer thicket," a few miles from the small town of Ezzell (pronounced "easel"). He recalls "underbrush so thick that at places you could not crawl through it." Henry grew up with no plumbing and no electricity. Nights ended strictly at 8:30 p.m. because Pohl senior worried over the nickel per gallon they spent on kerosene. On some winter mornings, when Henry rose to start the kitchen stove, he found the family's tub of water frozen solid.

Henry and his brother worked alongside their father: bringing up gallon after gallon of water from the well; milking the cows; carrying cans of milk-soaked corn to the turkeys; and gathering chicken eggs. He ranks these days as the best of his long life. "I have often thought of how nice it would have been," he says, "if my children could have had the opportunity to live one year like we lived between the years of 1940 and 1950." Pohl has maintained a deep Texas cadence and twang. He speaks in no particular rush, choosing his words like so many engineering decisions, with care.

Shortly after the summer of 1945, Pohl lived one of his proudest moments, bringing home a marvel: the family's first tractor. He carried a signed check to town and then rode the tractor back to the summer's thicket. "It was about forty degrees out, and it was raining. I was terribly cold and wet," he says now. "But I was the happiest boy on this Earth." The tractor would not provide nonstop delight, however. As Henry worked the family's land with this new machine, the steel steering wheel vibrated so violently that his hands would be bruised and swollen by the end of each day.

Within a few years of the war's end, not only would high-schooler Henry have built, from scratch, a sturdy barn for his family but he would also have connected their old home to the marvel of electricity. Vast swaths of rural America went without electric power until the mid-twentieth century. Henry used a Sears, Roebuck & Co. manual on wiring; it just made good, logical sense to him, and the Pohl home was one of the few to pass electrical inspection at the time. Henry was better than the professionals, a building inspector told him. But all Henry wanted was his own farm, and he planned to start in earnest after he finished high school.[11]

Young engineering minds blossomed everywhere as electric devices and new machines worked their way into all aspects of American life. In the small town of Elmira, New York, Marlowe Cassetti lived a less agricultural version of childhood, as a grandson of Italian immigrants. In August of 1945, Marlowe, not quite a teenager, was playing with one of his model planes when a friend rode up on a bicycle. "Did you hear the news?" the friend said. "We dropped an atom bomb on Japan." Neither boy really understood what the weapon entailed, but the war was soon over. And on "VJ Day," (victory over Japan), Marlowe's father gathered the family in their car and said they must go experience the pandemonium. Once in town, cars were everywhere, honking. A grown man sat on the hood of one, beating a little snare drum over and over, grinning ear-to-ear. "It was a fantastic relief," Cassetti says of the celebration. "It was kind of depressing as a kid during the war."

His father, Dr. Angelo Cassetti, had escaped crushing poverty in New York City's slums in the early twentieth century. Angelo's immigrant parents had told their bright boy there would be no more schooling for him—he needed to push a fish cart and bring in money—so he ran away from home and read every book he could find. By 1945, Dr. Cassetti worked in upstate New York, making house calls throughout a broad, hilly region, where doctors were scarce. Now a father of three, he maintained this passion for reading. When he returned to New York City, about once a year, it was never to visit his parents but to buy used books with his kids. "He could buy a trunk full of books for fifty cents each," son Marlowe recalls. More often than not, Dr. Cassetti spent evenings returning to work after dinner, but on those precious nights the doctor stayed home, Marlowe recalls reading magazines with his father, including *True: The Men's Magazine*, featuring fantastic stories of past lives and flying saucers.

As with many eventual NASA engineers, Marlowe was fascinated by flight. He recalls the grim Christmas holidays of 1941, after Pearl Harbor shocked the nation. His aunt brought grade-schooler Marlowe something amazing in that bone-chilling December: a model airplane kit. He carefully assembled the balsa wood parts. "I'd never done anything like that," he says. He wound up the propeller and stretched its coiled, rubber band engine. "I [would] go to one end of the kitchen, let it go from the linoleum floor, and it would fly off," he remembers. "It was a life-changing experience." Cassetti's voice retains a hint of a New York accent. He speaks quickly when excited about a story, moving nimbly between delight and wry analysis.

He went on to assemble larger, gas-powered versions and helped organize a model airplane club. He gathered ten or fifteen junior-high kids, and they even had special T-shirts printed in town. "All walks of life," he says. "Different economic classes, races, and everything else." (See Figure 1.3.)

He struggled mightily with his first gasoline-powered model airplane. "I unrolled the large set of construction plans and started to build the fuselage from sticks of balsa and bass wood," he says. "After many days, I realized that I was in over my head. . . . To add insult to injury, I couldn't get the engine to run beyond some occasional pops and backfires." An older neighbor boy (after overcoming a racial

FIGURE 1.3 Marlowe Cassetti, with one of his model airplanes, at age thirteen. (Photograph courtesy Marlowe Cassetti.)

reluctance to set foot in an Italian's home) eventually helped Marlowe get it going. Cassetti points to that as the start of his life as an engineer.

Marlowe quickly cultivated a reputation. His junior high school yearbook listed prophesies for students. Marlowe Cassetti, it claimed, "will be happily sending rockets to the Moon."[12]

Worldwide fascination with rockets had grown since the early 1920s. News outlets breathlessly reported physicist Robert Goddard's suggestion that self-guided rockets might one day visit the Moon. And in Germany, Hermann Oberth's writing dared to suggest that the technology could transport *people* into the cosmos; the first printing of *The Rocket into Interplanetary Space* sold out quickly, and subsequent editions spread across Europe. Science fiction stories, books, and movies adopted rocket travel as a staple. And, by 1930, filmmaker Fritz Lang had even introduced a dramatic element for starting a space voyage—the *countdown*—in his silent film *Frau im Mond* (Woman in the Moon).[13]

Circa 1945, a less central member of our initial cast, another twelve-year-old boy, enjoyed innocent visions of space travel in a mercifully dark movie theater, his sanctuary from Louisiana's sticky heat. Bobbie Brown—my father—marveled at a sleek rocket weaving a black-and-white path across the screen, guided by strings and sounding like an angry electric fan.

Bobbie was the youngest of five boys. Near their home in New Iberia, their father had worked the oil fields, shoveling coal into drilling rigs, as one fossil fuel sought to release another. Often in hand-me-down overalls, Bobbie roamed the dirt roads and swamps with his dog and best buddy, Susan. He developed what he now calls a "hungry litter appetite" as the smallest puppy at a table of older dogs. Dinner was often little more than rice and beans. As they moved about, from one oil town to the next, home sometimes had a dirt floor.

Even in rural Louisiana, they had seen constant signs of war. On the coast, pulling down blackout shades was not just a patriotic drill. Surrounded by a chorus of insects and frogs, Bobbie witnessed ghostly fires spreading on the Gulf of Mexico. German U-boats regularly torpedoed oil tankers and freighters leaving the Mississippi River. And in the streets, not even the few families with money drove new cars—the nation only built tanks and ships. Old beaters bumped through lumpy roads with stacks of cheap tires strapped to their roofs; any quality rubber went to jeep and airplane wheels, so the stateside tubes routinely burst. And in the Brown home, they prayed for the two older boys—one in the Air Force and the other in the Navy—to come home safe, while the three younger sons awaited whatever their mother could cook up with rationed staples.

When they weren't snagging frogs, crawdads, or fish from the bayous, Bobbie and his friends loved to lose themselves in movies. By collecting pieces of scrap metal, especially copper when they could find it, they earned pocket change from Uncle Sam. In assisting the war effort, the boys could parlay a sack of scraps into a double feature.

Interspersed with dramatic newsreels from Europe and the Pacific, westerns traded screen time with installments of Bobbie's favorite serial, "Flash Gordon." With the scientist copilot Dr. Zarkov, Flash used a submarine-inspired periscope to view distant worlds, enemy space ships, and the strangely cloudy cosmos. The serial's special effects appear laughable now, but they enthralled audiences at the time. "I remember the smoke from the ship's rocket engine would go upward, and that didn't bother anyone," my father says. "But there is no 'up' in space. We had a lot to learn." At the time, his dreams didn't rise much above a life with regular meals and a better fitting pair of shoes. Maybe in his boldest moments, he pictured a grown-up life that wouldn't involve shoveling coal. Working on a rocket ship was just something for the movies.

In retrospect, our story's cast could leave an audience shaking their heads. How could men from such different backgrounds, with such divergent dreams, find themselves knitting their brows and rolling up their sleeves together on behalf of a young president, hatching plans that lived somewhere between heresy and lunacy? It started just twelve years after the war, when their attention snapped to an alarming Cold

War surprise. Pohl, Cassetti, and Brown, with thousands of other Americans, many the first of their families to leave swamps, farms, factories, and mines, would gather in response. They followed a path set by a handful of visionary designers, including Faget and the transplanted von Braun. Together, these were the men and women for whom a Moon landing would become the unlikely last step in a long series of problems solved, calculations perfected, and gadgets tweaked.

2

1957—PATHS, POWER, AND PANIC

They both cried out. They both stood. The chairs toppled back, fell flat on the lawn. . . . They saw the brightening color in the sky and, ten seconds later, the great uprising comet burn the air, put out the stars, and rush away in fire flight to become another star in the returning profusion of the Milky Way. . . . Staring up, they heard themselves sobbing and crying.
—Ray Bradbury, *The End of the Beginning*, 1956

In just over a decade, Wernher von Braun had become a popular national figure. In retrospect, his climb from a mothballed artifact to a dashing national space darling could not have been more remarkable. But it fit his lifelong pattern of never accepting obstacles. His father recalled "absolute futility" in parenting Wernher. "Nothing worked. Any attempts to admonish him, or convince him of the inappropriateness of certain action, ran off not only like a drop of water, but like a drop of mercury, without leaving the faintest trace."[1]

After World War II, the United States had wanted to wind down its war machine, and the government viewed new missiles as an extravagance. Von Braun and his imported colleagues sat disillusioned in the American desert. Among his military supervisors, nobody seemed to care about his rockets. He was left to just reassemble the leftover scraps of old V-2's and watch them, more often than not, flame out, fall, and break apart over the sand, all while his new bosses shrugged. He submitted multiple resignation letters, but received only smug chuckles in response. When given the chance, like an invitation from the El Paso Rotary club, he practiced his English, buttoned his suit, and held forth. He spoke of rocket ships, space stations, and journeys to the Moon and other planets. In these years, von Braun determined he would take his case directly to the American people.

After the Soviet Union tested its first atomic weapon and communists took control of China in 1949, America hastily returned to missiles. They moved von Braun and his countrymen to an army base in Alabama, where he worked in the Redstone Arsenal's ballistic missile program. The Germans felt a sort of double salvation. Not only could they resume rocketry work in earnest, but in Huntsville they also returned to a life with greenery and forests, reminding them more of their homeland.

FIGURE 2.1 Walt Disney visits Wernher von Braun in Huntsville in 1954. (NASA photograph.)

Meanwhile, von Braun pushed himself further into the public (see Figure 2.1). He encountered some early speed bumps in Alabama. After a presentation to the Huntsville Kiwanis Club, a farmer said he would fly to Washington, D.C., on a bale of cotton before a man landed on the Moon. Soon after, von Braun received a letter from a woman advising him to stop his heretical talk of outer space—he should "stay home and watch television like the Lord intended!" But his confidence and charm made von Braun a natural scientific communicator, and he was relentless in his campaign. He made leading appearances in programs like Walt Disney's television special, "Trip around the Moon." Perfectly coifed and delivering impeccable, accented English, von Braun calmly described what a trip to our nearest neighbor would require. And he began writing for the general public as well, such as "Crossing the Last Frontier" for *Collier's*, "Cosmic Wonders" in *New York Times Magazine*, and "First Men to the Moon" in *This Week Magazine*. All he needed was national will with a generous allowance, and he would harvest the wonders of space for America. Wernher von Braun's Disney TV programs, in particular, mesmerized a number of eventual NASA engineers. "This show, probably more than anything else, influenced me to study aerospace engineering," said one. "And this wasn't the ordinary thing to do for a boy raised in a small Iowa farming community in the fifties."[i]

For some time, von Braun had wanted to use one of his rockets to put a satellite in orbit. In 1956, his team's latest creation, the Jupiter rocket, set records by reaching an altitude of 682 miles above Earth and covering a total distance of 3,400 miles. But the military kept holding him back. By restricting the heights and durations of his launches, they made sure he didn't "accidentally" put something in orbit around Earth. Through some clever labeling, he kept one of his best rockets intact and ready for future use, officially calling this a "long term storage test" of rocket hardware.[2]

He was not afraid to use fear and conflict to further his ambitions—whatever would nudge America spaceward. In fact, he had been warning the U.S. government that the Soviets would beat them to a satellite. Earlier in the 1950s, even while pushing a benign vision of space to the public, he privately advised army brass to consider a militarized space station and satellites armed with atomic weapons. And just a year before Sputnik, von Braun had promoted the great prospects for spying from orbit. His words also showcased his incredible ability to convey technical principles in lucid language. "The atmosphere is much more transparent from without than from underneath," he explained. "Pick up a piece of wax paper. Hold it close in front of your face and you see only a blur [like looking spaceward from Earth]. Hold it on a piece of newsprint, and it is perfectly transparent [like looking at Earth from orbit]."[3]

We can glimpse part of von Braun's operation through the early career of Henry Pohl, who started there as an enlisted GI in the fall of 1956. Henry was a newly minted engineer from Texas A&M University. His father had tried to talk his eldest son out of engineering, telling him he'd never be his own boss that way, and that he'd never have time for cattle. The elder Pohl pushed for law or medicine. Henry tried farming after high school, but a bad drought and lack of funds closed him down. He reasoned that, since he liked numbers and machines anyway, he could work a few years after college as an engineer, sock away some money, and then start a proper cattle farm. Then the army drafted him just as he completed his college degree.

The clarity and precision of Pohl's memories shine here. He knew his dog tag number for life after reading it twice, while some GIs never could perfectly remember theirs. He can recall driving to Huntsville (only his second time leaving Texas) in his 1953 Ford, and he writes of arriving there at 9:00 p.m., in a "light rain." When he

---

[i] This boy from Iowa was eventual NASA engineer Steve Bales.

met with his new commanding officer, he remembers it being "the first office on the right."

The Redstone Arsenal gave the enlisted men some choice as to where they worked. Pohl's commanding officer said that whatever Henry decided, he should avoid the "Test Lab," because the GIs working there got behind on their soldier training. Test Lab people were known to skip lunch, and they even worked weekends sometimes— it was awful, the officer told him. But none of that concerned Henry; far from home, he had little else to occupy him.

He went to see the Test Lab people and they said there was plenty of work for him if he was interested, but he had to be dedicated. After his interview, they told him to hold tight, skip lunch, and watch a noontime rocket test. "That dadgum thing looked pretty simple," he later said of the rocket on the test stand. The engine in the bottom didn't look much bigger than the one on his tractor back home, and it was just a long tube on top of that. They took him to the safety of a fortified blockhouse. Pohl watched them prepare and chatter back and forth over a loudspeaker. This was to be a "static" test—the rocket engine would be held in place and monitored for performance. "All at once that thing lit off," he recalls. "I had never seen power like that. . . . That beautiful white flame came out of the engine that looked like a gigantic cutting torch." What amazed Henry was that something the size of his tractor engine—which was impressive already—could weigh less but generate hundreds of times *more* power. He was hooked.

At first Henry followed the director of the static test stand around with a clipboard and simply did what he was told. After two weeks of that, he showed up for a day's test, but the director did not. The engineers grew nervous because the super-cold liquefied oxygen would start freezing parts of the engine if they didn't run it soon. Henry telephoned around the base asking about the test director and finally got the news. "Didn't they tell you?" a voice barked at him. "The test stand is yours. Run the goddamn test." Henry put the phone down, barely out of college and now in charge of one of the most powerful and dangerous engines on the planet. "That was the hardest 150 seconds of my life," he says now. After acting the part of a leader that day, he went home with a bad migraine. But each day became more comfortable than the last, and he craved the work itself. The same spirit that would solve any problem on the Pohl family farm would solve those that bedeviled rocket engines.

By the fall of 1957, Pohl answered to a German boss, a colorful and demanding figure named Guenther Haukohl. He had worn many hats in Germany, from flying experimental jet aircraft for the Luftwaffe to helping design underground rocket production lines serving von Braun's team.

One night in September of that year, Haukohl came into Pohl's office looking beaten and almost physically ill. Henry asked him what was wrong, and his boss said that the best American intelligence had confirmed it: The Russians were going to launch a major rocket and put something in orbit around Earth. Pohl had never seen the man look so depressed. "Henry," Haukohl said, "it's going to be very bad for the United States."[4]

Following World War II, the Soviets had conscripted a number of German rocket scientists who'd not fled with von Braun, and together they had assembled the remnant pieces of V-2 rockets. Stalin's oppressive regime had even allowed one Sergei Pavlovich Korolev, a brilliant Soviet rocket scientist, to briefly exit his life of Gulag confinement and take notes from the Germans at von Braun's old research center on the Baltic coast.

Nikita Khrushchev, shortly after taking the reins of the Soviet Union in 1953, discovered the rocket program that Stalin had hidden from even the most senior members of the party. And he discovered Korolev. When Khrushchev and his colleagues first saw one of Korolev's rockets, "we gawked at what he showed us as if we were sheep seeing a new gate for the first time. . . . We walked around the rocket, touching it, tapping it . . . we did everything but lick it to see how it tasted."[5] While Khrushchev granted the rocket scientist new freedom and relative comfort, he maintained a leash as well. Korolev would still toil in anonymity, hailed as the "Chief Designer" but with his name and face hidden from public. Khrushchev pivoted from the ways of his predecessors and branded the new Soviet path forward as technological. What need would the Motherland have for antiquated, standing armies (where he happened to have many political enemies) when they could instead paint a clear nightmare for the West with a web of missile trails encircling the globe?

Sergei Korolev's history-shaping masterpiece was to rocketry what Eastern bloc athletes would be to Olympic weightlifting: short, compact, and incredibly powerful. The thick R-7 rocket, only three times taller than it was wide, bundled together twenty rocket engines to create an unprecedented lift. The first edition, in the spring of 1957, burst into a fireball during an attempted launch. But Korolev was a master problem solver, driven not only by an urge to reach the heavens but also by the schemes of his boss. Korolev built a small hut exactly halfway between his team's rural launch pad and their nearby rocket assembly workshop. Coworkers said he slept little, trotting in alternate directions to check on one and then the other. After his partial liberation, progress was so steady that Khrushchev allowed Korolev to meet with some leading Soviet astronomers and to even start preliminary calculations for flights to the Moon and back. By early August of 1957, Korolev had worked out the kinks of the R-7 monster, and his team happily watched their squatty pride roar aloft and arc out of sight. On August 27, the Soviet Union announced to the world that it had built the world's first true intercontinental ballistic missile, capable of delivering nuclear weapons across oceans. There was little international response. On September 17, a follow-up announcement proclaimed they would launch Earth's first artificial *satellite*. The world collectively shrugged, assuming this fit a pattern of empty propaganda.[6]

Korolev had heard that the Americans and von Braun were talking about such a feat, and he knew the R-7, designed to carry a heavy nuclear weapon, should have little problem taking something much lighter than a bomb high enough and fast enough to achieve an orbit around Earth. Once he had the green light from party leaders, his team had designed a very simple device. At just 184 pounds, it was little more than a carefully enclosed pair of radio transmitters. They sealed this modest metal sphere with air inside to protect its electronics from the full vacuum of space. The radio would transmit beeps back to Earth. By varying slightly in length and timing, these pings would report on the health of the sphere's internal pressure. Soviet engineers were also worried about temperature—the first device to abandon the protection of our atmosphere might sizzle to death in the full rays of the sun. Korolev's team maniacally polished the metal exterior, hoping to reflect as much of the sun's heat as possible. And so, the first Sputnik was born. Awaiting another R-7 liftoff, the Soviet Union announced the device's radio frequency and invited the world to tune in.[7]

After working through many technical delays, Korolev watched his latest R-7 light up the night sky of October 4, and once it delicately placed its *Sputnik* in a large oblong orbit, Korolev gave a speech to his team. The Soviet Union had lifted the first human-made object to take laps about Earth—their own tiny moon. "Today the dreams of the best sons of mankind have come true," he said. "The assault on space has begun."[8] Khrushchev, informed by phone, nodded at the news. He congratulated Korolev and the engineers on another successful launch. Witnesses report that he returned to an ongoing state dinner in Kiev, glowing with the good news. But when he shared it with the assembled guests, they only smiled politely, having no idea what a "satellite" meant. Why should he have expected anything but another global yawn for their new announcement?[9]

The world's reaction to Sputnik was, of course, overwhelming, surpassing even the most hysterical dreams Khrushchev may have enjoyed that night. Americans in particular seemed to tremble with a new collective phobia. U.S. Senate majority leader Lyndon Baines Johnson took his dinner companions for a walk along the Pedernales River in Texas, looking upward for Sputnik. He said the sky now "seemed almost alien." And two weeks after the launch, *Life* magazine tidily summarized the nation's response. "Let us not pretend that Sputnik is anything but a defeat for the United States." The only path to redemption, according to most, including President Eisenhower, involved a new dedication to math and science.[10]

It's easy now to say what Eisenhower tried to relay, in vain, at the time: The public was overreacting. As he told an advisor, "One small ball in the air . . . does not raise my apprehensions—not one iota." But the Soviets had aggressively claimed Eastern Europe after the war and then had laid siege to Berlin. In Asia, communism had already enveloped a huge swath of the world's people. In this light and in the ominous glow of ever larger nuclear tests, the official Soviet Sputnik announcement promising "larger and heavier" satellites sounded grave. And Americans couldn't help but hear a challenge in the Soviets' concluding statement of October 5: "Artificial Earth satellites will pave the way to interplanetary travel."[11] They had just claimed Earth orbit, and tomorrow they would take the solar system!

Younger Americans were, in many cases, less shaken than their parents. Their minds had ample room for fascination. A number of eventual NASA engineers changed their college majors within days of hearing the Sputnik news, in part for patriotic duty but also for the exciting lure of space. And some even younger Americans took to rocket building.

We get a detailed glimpse of this through the stories of eventual NASA engineer Bob Austin, the son of a Tennessee dairy farmer. He attended high school in a small town north of Nashville, "just before you crash into Kentucky." His father wanted him to eventually take over the family farm, but Bob, after watching a family friend build a radio from scratch, had his mind set on a career in the magical new world of electronics.

After Sputnik launched, American newspapers published the satellite's regular North American transits and its transmission frequency. The boy and his friends got to work. "We set up our receiver and a tape recorder and recorded the beep beep beep of Sputnik when it came over." After one of their teachers learned what they had done, the high school called for an assembly. An excited but unusually quiet auditorium full of students and teachers waited until the boys on stage hit "play" and Sputnik's monotonous electronic pings marked the time. With nearly two beeps per second, it thumped like a sprinter's heartbeat, predating techno music by a half-century. They let it run. "We might as well have been Einstein," he says of their classmates' awe.

Meanwhile, as with so many boys around the country—it was nearly unthinkable that their sisters would join them in this era—Bob and his friends had started building their own rockets. Model kits weren't yet available, but enthusiasts could easily buy the fixings for primitive rocket fuel. "Back then," Austin says, "you could go to the drug store, buy your carbon, buy your potassium, buy your sulfur." They also needed gunpowder, ground fine, to keep everything burning quickly enough. A friend's father worked for the FBI and helped the boys get a batch of shotgun shells, which they carefully emptied and sifted. They fashioned a rocket from an aluminum tube, filled it with their powder mix, and learned firsthand a fundamental truth of any rocket program. As Austin puts it, "There's a narrow line between exploding and a launch." In many ways, a successful launch is a controlled explosion happening at just the right speed.

After a number of loud failures with these powder-packed engines, they set out to make a hopefully more reliable *solid* fuel. They would carefully mix an explosive brew and let it harden. The young team started their recipe in Austin's grandmother's kitchen. "Bless her heart," he recalls, "she thought we were trying to shoot down *Sputnik*." Their first batch didn't go well; they'd talked with their physics teacher about the right recipe but apparently had missed a crucial line of directions. As Austin toiled at the stovetop, stirring the ingredients, he let the temperature drift too high and the rocket brew went off. One minute he was stirring, and the next minute there was smoke everywhere. The fuel had burned his hand and one side of his face. Despite the high ceilings of the kitchen, the boys fell to all fours to avoid the thick smoke and crawled outside.

This scene, not unlike those in Homer Hickam's *Rocket Boys*, played out in hundreds of towns. "So many people got hurt trying to make their own fuel that you could no longer go buy the components," Austin says. But they had a more local problem: his father. Unhappy with the accident and near maiming, Mr. Austin told the boys they could try just one and only one more launch, this time with adult supervision. With his help, the would-be rocketeers finally had a solid-fueled rocket ready to launch. They moved off to a safe distance and counted down. "That sucker blew up like a cannon," Austin says. And for all he knew, his days with rockets were done.[12]

Wernher von Braun's chief reaction to Sputnik was one of fury. He'd wanted his own rocket and his new country to be first. He had been thinking of a satellite for years already. A colleague recalled von Braun discussing the detailed steps of a satellite launch as early as 1952 and even fearing that the Russians would have the nerve and ability to do it first.[13]

A reporter informed Wernher von Braun of Sputnik's launch while they attended a cocktail party. In his consternation, von Braun marched directly to his army bosses and said they had to finally unleash him. "We can fire a satellite into orbit sixty days from the moment you give us the green light," he told them. Within a few weeks, he was speaking in Washington, D.C., saying everyone had failed to foresee the effects of "an omnipresent artificial moon," and he told the audience that surely the Soviets had more surprises to come. But the first crack at matching the Soviets was not given to von Braun's army rocket, the Jupiter, but instead to the navy's rocket, the Vanguard, already in advanced trials.

At an untrained glance, the navy's rocket and von Braun's army version were quite similar.[ii] Each was about seventy feet high and in the best circumstances their

---

[ii] The precise rocket names are not critical to our narrative here, and we will oversimplify rocketry's rich family tree. In the early weeks of the space age, missiles from three branches

handlers thought they could deliver a fifteen- or twenty-pound object into orbit around Earth. They each had three "stages," or segments.

A rocket's fight against gravity makes every pound count. To minimize carrying dead weight, rocket architects often plan to discard segments, or stages, during flight, as soon as they've been emptied of fuel. By ejecting an empty stage, the rocket loses the weight of that structure, its storage tanks, and even its engines, making the remaining journey much easier. A rocket could have fifteen or twenty stages, but then it would need that many separate sets of engineered fuel tanks, monitors, plumbing, and so forth. And each stage must be precisely ejected by "pyrotechnics," or little explosives that fire precisely to release the stage—still more places where engineering flaws could hide and doom the flight. In the end, rocketry requires a negotiation between fuel efficiency (more stages) and reliability (fewer stages). Three or four stages formed the best compromise for the early decades of the space program, from the Jupiter and the Vanguard all the way to von Braun's eventual triumph, the Saturn V.

As military branches lobbied for their separate rockets, Khrushchev delighted in America's panic. He used satellite success to emphasize a shift away from conventional arms. He said his countrymen must embrace the missile era, where they would lead all nations. And to rub salt into the fearful wounds of his enemies, he offered the American government his services if they wanted to put payloads into orbit.

He wanted more success, and he pushed Korolev for another milestone. In early November, the rocket scientist delivered and the USSR proudly announced a *second* satellite, a true whopper named Sputnik II. It placed over one thousand pounds of payload into orbit, and since it carried the shell of its last stage as well, it orbited as a six-ton monster. Sputnik had scared the public, but Sputnik II alarmed military analysts. The Soviets, by all evidence, could deliver something the size of a nuclear weapon into space or wherever they pleased. And hinting at future ambitions, they also included the first mammal in orbit, as Laika the dog survived the violent rocket launch, and the Soviet Union broadcast her heartbeat to the world. As the press dubbed this new milestone "Muttnik," the American public actually wondered if Laika might once again set four paws on Earth. But that was never in the Sputnik II flight plan; Laika died from overheating, after just a few hours in orbit. (Despite the loss of Laika, more dogs were on the way, some of whom survived a return to Earth, and Korolev was reportedly attached to his pioneering space dogs, grieving any loss.)[14]

As Khrushchev crowed that the Soviet Union would surpass the United States economically within fifteen years, *Life* magazine featured Wernher von Braun on its November cover. He told them of the many letters from youths that he answered,

---

of the military provided some of America's only space rocket options: for instance, Atlas (air force), Vanguard (navy), and von Braun's Jupiter (army).

always emphasizing the study of math and science. And with his audience firmly in mind, he relayed a different letter. "One lady wrote that God doesn't want man to leave the Earth and was willing to bet me $10 he wouldn't make it," he said. "I answered that as far as I knew, the Bible said nothing about space flight but it was clearly against gambling." He had largely won the nation over, and now the nation needed his help.

A few days after Sputnik II, the U.S. government gave von Braun permission to start preparing one of his Jupiter rockets for launch but pulled up short of a "go ahead" order. He had threatened to resign if they didn't let him move full speed ahead, and the army relented.[15] But the navy's Vanguard would still have the first opportunity to loft an American satellite into orbit—just a modest start, a three-pound device.

Non-military facilities started tinkering in rocketry as well. By 1957, Marlowe Cassetti had left rural New York and was a newly minted engineer from Georgia Tech. He accepted a job running tests at the government's aeronautical research labs in Langley, Virginia. After his earlier moment of rapture, with a flimsy model plane leaving his family's kitchen floor, he finally had his hands on the real thing, working with a wind tunnel to measure the way an airplane wing cut and molded the rushing air. He recalls being surrounded at Langley by flight veterans of World War II. On hearing the news of Sputnik "they were all very much shaken by it."[16]

One such veteran had found his way to Langley after his tours on the USS *Guavina*. On a lark, Max Faget and his friend Guy Thibodaux had followed a job lead from Louisiana to Langley, Virginia. "Max called me," Thibodaux recalled, "and he said his dad had a little car that had airplane tires on it at that particular time, because you couldn't get tires for civilian automobiles. His dad told him he could have the car, and we'd go look for a job."

The boys interviewed with a mild-mannered but earnest aircraft engineer named Robert Gilruth. Years later, Gilruth would still chuckle and shake his head at the memory. He recalled that the two Louisiana kids "were kind of grubby." Unshaven and wrinkled, they had been sleeping in Faget's father's car. Despite those first impressions, Gilruth would eventually call the day he interviewed and hired them the best day of his life. He quickly saw the zeal of the two young men, doing whatever needed to be done, from any engineering task to carrying boxes or brewing the morning coffee.[17]

Faget became a force at Langley, working on the design of hypersonic aircraft, planes that would have to withstand speeds greater than the speed of sound. "You know, when I joined [Langley] in 1946," Faget later said, "most of the people that came in . . . were young, very young people who had spent two, three, four years in the war. . . . I think the experience of the war gave everyone a sense of urgency . . . to get into jet-powered airplanes, get into supersonic airplanes and things like that." Langley, long a pioneer in aircraft research, would soon become the central hub of the nation's efforts to reach space.[18]

Despite a rough-edged confidence that didn't suffer fools, Faget was also never far from a smile and a laugh. (Authors Charles Murray and Katherine Bly Cox summarize his persona as a sort of "cheerful ruthlessness.") And colleagues quickly learned they had a dreamer in their midst. He once stood on a high balcony overlooking the main Langley workshop floor. He had taped together pairs of paper plates and stood earnestly tossing them around the shop to settle among the busy machinists and technicians. One coworker, less familiar with Faget's modus operandi, asked him what the heck he was doing. "I think these things will really fly," Max said of the flying disk model. "We have some lift over drag in this thing."[19]

Faget and his colleagues organized a national conference on hypersonic aircraft, set for October 1957. "A couple of weeks before that meeting, the Russians put up their Sputnik," he later said with a laugh. "Of course, that set us all back, you know. At that meeting the discussion then was . . . maybe we shouldn't try to fly up to those velocities with airplanes, but maybe we ought to bypass the airplane role and go directly into rocketry." But the discussions in late October of 1957 went further. "We talked about rocketing men up into orbital velocity and how to get them back."[20]

Soon after, Faget and a small group of his coworkers started taking mandatory night classes. They pivoted from things that flew at high speeds through the air to things that flew at even higher speeds *above* the air. They had to absorb the basics of space and astronomy, and there was no time to lose. They held classes in a local elementary school. One night they took a break when Sputnik was due to pass. As they tried to spy it racing overhead, some of them began to have hushed conversations about putting a man in orbit. Not only was there a lot to learn, there was even more that was wholly unknown, beyond what could be taught. This stuff wasn't in any textbook yet. In 1957, scientists had only speculative theories as to whether a person could survive in the weightlessness of space. What would happen to their organs, their blood vessels, and their brains without an "up" direction or the grounding of gravity?[21]

The engineers ramped up efforts on Virginia's nearby Wallops Island to study not only rocket launches but especially what would happen to a rocket cone as it zoomed back to Earth. "Wallops Island was a mosquito- and sand-fly-infested beach that had a bunch of wild ponies on it," Faget's friend Thibodaux recalled. "We used to have to shoo them off when we fired rockets." The engineers would live there for days at a time in a set of Quonset huts. After long days of work, Gilruth, Faget, Thibodaux, and the others would stay up into the evening trading issues of *Astounding Science Fiction*.[22]

On December 6, racing the Soviets and history's calendar, the navy's Vanguard rocket ignited and lurched upward from a Florida launch pad. With cameras rolling and broadcasters relaying their excitement, Vanguard suddenly stopped its ascent and slowly fell to the pad. Like a fainting heroine of black and white movies, it careened to one side. A fireball consumed the scene, as all the unused fuel and oxygen combined at once, blowing the tiny satellite clear of the inferno. Nestled in the nearby shrubbery, it assumed it must be in orbit and started beeping out its signals. As the nation's spirits sank, the press labeled the disaster "Kaputnik," "Stayputnik," and so on. The term "missile gap" emerged in Congress and in the newspapers, as did the notion of a "space race."[23] Surely the Soviets would next put a man in space, and eventually, a fleet of spaceships with unimaginable powers.

3

1960—SILENT MOVIES AND
OLD-WORLD EVENINGS

By the start of 1960, Wernher von Braun had become a sort of national hero. His rocket, a modified Jupiter, had redeemed America in early 1958, placing a small satellite, Explorer 1, into orbit around Earth. Within weeks, von Braun's face and (sanitized) life story graced the cover and pages of *Time* magazine. The following year, Notre Dame University honored him with their "Patriot of the Year" award. By the end of 1958, the military had given his team a green light on developing a new super-rocket, something much more powerful than the Jupiter series. And his outward ambition only grew; as the Soviets were now sending unmanned probes toward the Moon (their Luna series), von Braun manipulated his army bosses to the point that they drafted "Project Horizon," a preliminary plan for a manned lunar garrison.[1]

The town of Huntsville had blossomed. In one decade with von Braun and company building rockets, its population had quadrupled to seventy thousand. Formerly a cotton depot for northern Alabama, the town swelled with wave upon wave of scientists and engineers—it was morphing into "Rocket City." And some of the new citizens, the ones of German extraction, wanted more of the comforts of home. The Huntsville Symphony Orchestra (now Alabama's oldest and longest-running) emerged in 1955 with many of its forty members hailing from von Braun's ranks.

Young Marlowe Cassetti made regular work trips from Langley to Huntsville for coordinating the early launches. To his eyes, the rapid growth never disguised the town's roots. "Can you imagine that place in the late fifties and early sixties? It was really a cow town kind of thing," he says now. There were few options for lodging, and he usually stayed in an antiquated hotel downtown, a place with huge gaps under the transom-topped doors, awkward plumbing bolted to rooms as an afterthought, and no air conditioning to fight the summer heat. "It reminded you of a place from the 1930s. . . . I used to call it the Erskine Caldwell after the southern writer." (Caldwell, author of *Tobacco Road* and other novels and short stories, highlighted social dysfunction in the American South.) But Cassetti appreciated that his government allotment of $16 per day went a long way in Huntsville, laying down a handful of change for a generous southern breakfast of ham, grits, and eggs.[2]

By 1960, Cassetti, Pohl, Faget, and von Braun had joined forces under a new banner. The National Aeronautics and Space Administration (NASA) emerged from the Congressional Space Act of 1958, a symptom of the national post-Sputnik panic. NASA owed its basic DNA to an older organization, Langley's National Advisory Committee for Aeronautics. (NACA had also formed in response to breathtaking technology, the first flights of the Wright brothers.)[3] Cassetti notes that, from the start, none of the engineers pronounced the acronym: "I would use the analogy of the FBI," he said. They would just spell it out, pronouncing each letter: N, A, S, A. The first he'd heard it spoken as "nasuh," "nayza," or "nasaw" was in a reporter's radio broadcast. Whether pronounced or spelled, NASA quickly began absorbing relevant government laboratories around the nation, and von Braun's army ballistic missile group looked like an essential piece. Despite the army's reluctance, von Braun and company joined NASA in 1959, and von Braun, at the age of forty-eight, had his first non-military post since his college days. In a last nod to the army, Eisenhower agreed in 1960 to name von Braun's own outfit after a recently departed five-star army general, George Marshall.[4]

Wernher von Braun's prestige within the U.S. government and his national profile continued to grow in 1960. A sentimental biopic, *I Aim at the Stars*, hit theaters that year, and a common joke emerged, again referencing his former employment: "I aim at the stars, but sometimes I hit London."[5]

Despite his excitement as America joined a space race, von Braun and his colleagues faced enormous challenges. His chief scientific advisor, Ernst Stuhlinger, set out some of the obstacles to space exploration in a clinical list. When first briefing their new NASA bosses from Washington, D.C., Stuhlinger said that, even ignoring the difficulty of lifting people into orbit, possible spacefaring ships would face many hazards that were still completely uncharted. While the atmosphere shields us from a great deal of radiation flying through space, a ship would lack such protection. What would space radiation, including all sorts of heavy, fast-moving particles, do to the electronics and materials of the spaceship, to say nothing of any humans inside? How often would a ship collide with asteroids, even tiny pebbles, moving at orbital speeds? Furthermore, how could a ship withstand the absurd temperature extremes in space? One side, facing the sun, would automatically start roasting at 250° Fahrenheit, while the same ship's backside, facing the darker depths of space, would freeze to −250° Fahrenheit. And the weightlessness expected for a ship in orbit posed its own mysteries. Many mechanical systems and engines use fluids, and gravity helps us control them. We always know where the fluid will be: at or close to the bottom of its tank. But in orbit, away from Earth, fuel could float around wherever it wanted within its container; how would we convince it to move toward a specific valve and onward to an engine? Moving from humans and fluids, Stuhlinger listed an even more daunting set of material problems. In the vacuum of space, sealants would start evaporating around doors, windows, and seams. Various metals would violently expand and contract their sizes as they moved from the scalding, sunny side of Earth to the frigid, shaded side. He saw deficiencies everywhere.[6]

Even in Huntsville's earthly laboratories, problems with materials multiplied. To build ever more powerful engines, von Braun's teams worked to perfect turbopumps that had no function other than rushing fuel and oxygen toward the blast chamber and the controlled explosion providing the rocket's power. But these turbopumps had to withstand unprecedented extremes of temperature, including the yin of liquefied oxygen mere inches from the yang of rocket fire. By the end of 1960, the team in Huntsville had suffered eleven dramatic pump failures as materials wilted during tests, leaving fuel and oxygen mixing in uncontrolled passion for one another. Violent explosions punctuated a number of these dead ends.

But the Germans brushed the soot off and worked in the same logical stepwise approach they had used in World War II. A few new rocket engines did survive their "static" test phase (bolted in place). The first test of a new "Saturn" type of rocket stage—the type we'd need to go to the Moon—took place in the summer of 1960; it had roughly fifteen times more power than von Braun's Jupiter series. And now the newly named Marshall Space Flight Center had a wholly different kind of impact on Huntsville and Alabama: a sonic wallop. The first Saturn model strapped eight rocket engines together. Skeptical engineers had labeled the concept of gathered engines "cluster's last stand."[7] When the eight simultaneously roared to life, deep sound waves rushed outward in all directions. Engineers of that era talk of meetings suspended during the engine tests, watching the seconds tick past as their pencils vibrated on conference tables. Henry Pohl recalled engineers staggering about the test site and retching after an early Saturn test. Even for the sound below the range of human hearing, engineers felt the vibrations in their stomachs and ribs.

Soon, townspeople for miles around reported similar effects. It felt like an earthquake: plaster chunks fell from walls; windows rattled and broke. All sorts of complaints made their way to the center from homes and businesses up to fifty miles away. A team of engineers took over both investigating these claims (e.g., "You say it made your *porch fall off*?") and devising tests to minimize future damage. Near the test stand, they erected horns several feet wide, and they equipped a couple of vans with sensitive recording equipment. Before any Saturn rocket test, the horns would emit a low-toned, obnoxious squawk, and the vans, moving about town, would report the day's sound results in various neighborhoods. The day-to-day difference was striking, as certain weather and cloud patterns reflected sonic blasts back to Earth with great swings of intensity. The squawking program helped the engineers call off certain tests and avoid most damage to the town. And citizens begrudgingly preferred the frequent honking to shattered glass.

The engineers devised careful sonic measurements *near* the rocket as well. They set up little orchards of carefully spaced microphones in the lawn near the test stand. In this way, they monitored exactly which directions were getting the loudest sounds.[8]

It was an earlier, smaller type of engine test that had snared young Henry Pohl and changed his life. By 1960, Pohl's attention to detail and inexhaustible curiosity continued to accumulate responsibility. He recalls a typical Saturday night, still turning wrenches at 11:00 p.m. "We had a [liquid oxygen] valve froze up, and trying to get it thawed out, when one of the technicians came to me and says, 'Henry, you ought not be up here tonight. You ought to be in downtown Huntsville having a good time.'" Pohl paused, thinking but not replying. "And I really couldn't sit there and explain to him that there was nothing in this world that I wanted to do worse than what I was doing then. I was up there that Saturday night because I enjoyed it. . . . I was hooked on those rocket engines."

At about this time, Pohl met and started dating a school teacher named Helen. Years later, she liked to tell their children about their first date, an outing so dull and nerdy that it almost didn't lead to a second date. Henry took her—where else?—on a tour of the Marshall Space Flight Center.

Pohl had just completed a special project for von Braun, building a one-twentieth scale model of a Saturn rocket's first, most powerful stage. It was more than just a proof of principle. Such a model let the engineers study possibly catastrophic problems in great detail, with a fraction of the risk and expense. "So I got the job of developing some model rockets, small ones, and clustering eight of those little rockets in a thirteen-inch circle diameter," Pohl said later. "I'd never designed a rocket engine, didn't know much about them. . . . I checked out all the books that I could find. . . . I'd stay up 'til two o'clock in the morning studying those things and playing with them." With this scale model, the engineers could test the entire first stage of a Saturn in any environment they wanted, including in a vacuum (simulating high altitude or space), or in a wind tunnel (simulating a rocket punching through the lower atmosphere).

His model of eight clustered little engines eventually worked flawlessly, and with a number of tiny sensors placed throughout the model, the Huntsville team came to understand the spread of heat through a rocket's business end. They then used Pohl's model to examine a crucial challenge: deflecting a rocket's output, be it on a test stand or in an actual launch. Several early Saturn rockets had destroyed themselves with their own exhaust. When an engine's super-hot leftovers linger or return to the blast chamber, they can overheat and rupture the entire engine.

Contrary to what a layperson might expect (and contrary to what several leading newspapers had published in the mid-twentieth century), the billowing exhaust of a rocket does not functionally "push" against a rocket pad to provide lift. If that were true, the engine wouldn't lift a rocket once it was airborne, to say nothing of firing in space, with nothing to push against. Instead, the rocket is literally throwing pieces of itself in one direction, so the rest of it will necessarily move in the opposite direction. Picture a child on ice skates, throwing rocks at one side of a lake only to find herself sliding toward the other.

Given that a rocket doesn't need its exhaust to push against a launch pad, the pad usually includes a large hollow underneath, aiming to keep the exhaust from disastrously fouling the engines. For smaller rockets, a deep cavity is good enough. For the types of engines in the enormous Saturn series, there was still trouble brewing, as Henry Pohl's model discovered. A normal, flat-bottomed cavity (even a deep one) would just bounce the exhaust right back into the engine. Henry and his colleagues came up with a solution, and it was not unlike a plow blade, with a long curve on either side, splitting the exhaust and gently ushering it out to side vents underneath the launch pad. This worked well for Pohl's model, and NASA eventually incorporated this solution for their launch facilities in Florida.

The rocket model turned out to be a showpiece for Wernher von Braun, and when he ordered a short film summarizing the project, Pohl became a movie producer. "So I took mostly test film, and I'd go home at night and I'd sit there," he said, "and I'd spool it back and forth and . . . back then you'd cut and glue it together." He took a voice recorder and narrated the film and handed it over to his bosses. They liked it, but "of course, they didn't like my voice, the way I talked." Henry still has a copy of this film, in its finished form with tell-tale 1959 "I'm narrating a serious project" voice. Missing is Henry's cayenne-and-vinegar south Texas twang. But we get to see young Henry, slender and intense, assembling his model. The plumbing for the engines looks like the intertwined roots of an ancient tree, and you wonder if this thing could actually work until they light it up and all eight fire in perfect unison.[9] (See Figure 3.1.)

At about this time, Bobbie Brown, having changed his named officially to Robert Brown, took a new job in Florida. As so many Americans did in the mid-twentieth century, my father used the GI Bill as a climbing rope from poverty. He'd enlisted in the navy during the Korean War. Far from the battle lines of the Korean peninsula, he learned the world of electronics, using sonar to track and chase Soviet submarines in tense games of cat and mouse around the Atlantic and the Gulf of Mexico. After his four-year term, he used the GI Bill to complete a physics degree back home in Louisiana, and by 1960 he was fine-tuning missiles for government contractor

FIGURE 3.1 Henry Pohl in a 1959 NASA film detailing a small-scale working model of a Saturn rocket. (Still taken from footage provided by Henry Pohl.)

Honeywell. My mother still tells a story of their first hurricane season near Tampa Bay, Florida, where storm winds blew rain through the cinderblock walls of their small rental.

Robert liked the work, despite its old-school industrial setting. He and the other young engineers worked all day at their desks and tables in a large open room, with supervisors a floor above, literally overseeing the work through a glass window. He was tasked with "inertial guidance," a big topic for missiles and rockets alike, starting with von Braun and moving into true spaceships. Once an object leaves the ground, how does it know up from down and north from south? How can it tell where it is or how far it has traveled? In particular, how can a rocket discern any of this without eyes?

Remote control, with ground-based engineers constantly detecting and sending information to the missile, including correctives like "a little to the left," becomes impractical for long-range flights. Instead, self-guided missiles use clever devices and a few calculations to orient themselves. The primary device is a gyroscope, simply a glorified spinning top. As a precisely machined metal sphere rotating rapidly within a well-oiled pocket, the gyroscope always maintains an "up" direction. Imagine a fast-spinning top on a cutting board; if we tilt the cutting board a bit, the top still points directly to the ceiling. Similarly, as a missile changes its direction, the gyroscope simply pivots in the oil, maintaining its original "up." (A spinning object always seeks to maintain its spin direction; this principle also helps a well-thrown football, or a rifle bullet, maintain its trajectory without tumbling.) By using a few gyroscopes, with each one pointing and spinning in a different direction, a missile (or a rocket) can always know up from down, left from right, and forward from backward. These become even more important once an object ventures into the completely disorienting realm of *space*. Another device benefiting a missile or spaceship is an "accelerometer." It detects how motion *changes* when an object speeds up or slows down. If such a device, placed in your car, carefully records each lurch, each speeding up, each turn, and each stop, it can know precisely how far you've traveled and in what direction. (Humans have their own inertial guidance devices at work in the inner ear; symptoms of vertigo, for instance, occur when the inner ear falls out of whack.)

One of my father's jobs was to make sure these guidance instruments would keep functioning at altitudes high enough to flirt with the edge of space. He recalls battling a leaking instrument box. To keep functioning at high altitude, the electronics needed a normal envelope of air, just like the original Sputnik, but no matter what he tried, air would start escaping the box when they put it in a vacuum chamber. Finally, in scrutinizing an access plate on top of the box, he removed a leaky rubber gasket and turned it upside down, completely against its designed orientation. "That fixed the leak," he says. Those particular instruments continued to fly in tests and launches for years to come, all with their gaskets upside down.

The rocket work of 1960 served two masters: national defense (trying to bridge the supposed "missile gap" with the Soviets) and the space race. And the Soviet Union showed no signs of slowing. Unknown outside Khrushchev's and Korolev's inner circles, they had faced some catastrophic failures as they pushed for further humiliations of America. In one case, Korolev's cluster of five rockets came unhinged after the central one exploded, sending the four supporting engines sketching their own random trails through the air, like snakes loosed from the head of Medusa. In another, the Soviets suffered what is still the deadliest rocket accident of any nation: the Nedelin disaster. In the fall of 1960, Khrushchev wanted a spectacular ballistic missile test to coincide with his famously bombastic visit to the United Nations in New York. The Soviets set Korolev aside for the moment, using a rival scientist's rocket that was capable of delivering heavier warheads, something to put a proper scare into the Americans. Shortly before launch, Soviet engineers detected something amiss and stopped the countdown. As technicians swarmed the launch pad, looking to find the problem, a fire burst forth on the second stage. Suddenly, an angry flood of burning rocket fuel cascaded onto the pad and the doomed technicians. Records released in contemporary times list a death toll of 165, including the supervising military official, Nedelin. The Soviet control of information hid these events for decades to come.

What the American public heard about the 1960 Soviet rocket program was one new accomplishment after another. Now the Russians could take live dogs (collected like true members of the proletariat, from the streets of Moscow), have them pant their way around Earth for an orbit or two, and then return safely to the ground, alive and ready for kibble. The Soviets also put a mannequin in orbit, hinting of things to come. It seemed that they'd soon have a comrade alive in space, looking down on the heartland of America, smirking at inferior engineers and helpless politicians.[10]

Such fears stoked America's already restless presidential election that year. *Life* magazine asked a series of questions about "our national purpose." The editors noted the roaring economy but asked, "What now shall Americans *do* with the greatness of their nation?" One presidential candidate emphasized that the Soviet menace might short-circuit the postwar success of America. Many citizens of Earth, John F. Kennedy said, "are not at all certain about which way the future lies," but they saw two clear paths, and the way of communism beckoned. "The first vehicle in outer space was called Sputnik, not Vanguard. . . . The first canine passengers in space who safely returned were named Strelka and Belka, not Rover or Fido, or even Checkers."[11] The latter pooch belonged to Kennedy's rival in the presidential race, Richard Nixon. Kennedy, who narrowly (to put it kindly) won the election that year, promised to challenge communism on all fronts, from Asia to South America, and from the sea floor to outer space.

And what progress could NASA claim by the end of 1960? At the agency's birth in 1958, Robert Gilruth, the long-time veteran of the Langley flight research program, brought together what he called a "Space Task Group": thirty-seven male engineers, and eight female technical secretaries, literally called "computers." Gilruth himself was the oldest person involved, just entering middle age. By the end of 1960, this group had grown to about six hundred people, and their first main task was to shoot a human being into space and retrieve him more or less alive.

Gilruth would guide a lot of NASA's manned spaceflight efforts through the end of the Moon missions in 1972. As Space Task Group member and eventual NASA center director Chris Kraft has written, "No man of space did more or received less credit than Robert R. Gilruth."[12] There are several key aspects worth noting here, at the ground floor of the NASA program. First, Gilruth favored administrators with all the grease stains, burns, and reflexes that came from building devices and fixing machines themselves as engineers. Next, Gilruth was excellent at hiring people who could do things he could not. Far from wanting to surround himself with the comfort of slightly inferior clones, he sought out smarter people (when available) and certainly people with skills he lacked. He was also incredibly persuasive, giving opposing points of view a full listen before calmly sharing his own. Gilruth told his troops in 1960 that "the most important" aspect of their work would be that "designs, procedures, and schedule must have the flexibility to absorb *a steady stream of change*" (emphasis added), as they learned more about "space problems."[13] He could not have been more prescient.

Though it may be difficult to believe today, when he started recruiting young talent to the Space Task Group, many engineers were leery of chaining themselves to the novel idea of space missions—older engineers warned younger ones that space would be a flash in the pan and a sure dead end for their careers. Cathy Osgood, one of the first women to join NASA's efforts, recalls sitting with her husband, trying to make the decision. He had a job offer from NASA, where he could help them establish a number of remote communications stations at strategic and hopefully friendly spots around the globe. (It was a wholly new problem, with any orbiting spaceship spending *most* of its time out of touch with North America.) But he also had an offer from the navy that would land him and his young family in sunny San Diego. "So what we did," she recalls, "we had a long legal pad and we made lists of each job, its pros and cons, and what it meant to us as a couple, of what it meant to our family, and what it meant to our future. It had to differ by ten percent for us to make a decision." How did they measure ten percent on a list like that? Osgood pauses and then speaks of her husband. "He was a very interesting person. He graduated from MIT in chemistry."

The couple went over and over their list, trying to quantify this big decision, but they could not reach the required ten-percent threshold. Then her husband's parents visited for the holidays, "and they started talking all about what was in the news about space," she says. "I hadn't paid attention to it, Sputnik and all that. I was busy and had kids to take care of." Her in-laws' enthusiasm tipped the balance. Her husband took the NASA job, based in Virginia, and Cathy joined the agency as well.

Osgood became one of the female computers, performing calculations equivalent to what we might do today with a spreadsheet program. "You'd multiply column one by column two and that sort of thing," Osgood says. "We were just crunching numbers, and then we had to plot it all by hand." Starting in the 1930s, Langley had employed groups of mathematically skilled women to support the computation-intensive grind of aerodynamics research. The need for their reliable math skills provided a unique opportunity for women at the time, though their work was sometimes overlooked. (And until 1958, the work was segregated as well, with separate wings for white and black computers.) Even inside Langley, many staff members passing by these wings thought the women were performing routine secretarial work.

Just as Osgood joined the all-female computer squad, the Space Task Group cautiously began using a new type of equipment, namely an *electronic* computer. Engineer Hal Beck recalls everyone gathering as they "oohed and aahed about the wonder of flashing lights and how fast it could add two numbers." But the humans were by no means out of their jobs. Early electronic computers were not especially reliable, and humans frequently double-checked and corrected the electronic calculations. In addition, an enormous early IBM machine could spare none of its scarce memory to display results. It would print out its raw lists of numbers, and the team of women computers would then turn that into displayed data, including hand-drawn plots of rocket paths, rocket speeds, and capsule orientations in space.[14]

I've asked a number of members of the Space Task Group about the relationship between men and women at Langley and between white and black employees. They emphasize witnessing and experiencing respect flowing in all directions, and, as one said, "The government already had quite a few rules for how you were supposed to treat one another" by 1960. But many engineers admit to hearing "vernacular" in the workplace, referencing an array of period terms that today range from uncomfortable to unacceptable. But they all return again and again to what united everyone: the intense focus on the work, on getting things exactly right. Langley—the roots of NASA—tended to celebrate intelligence in any package.

Langley employees recall mixing socially across gender and color lines in the mid-1950s. At the dawn of NASA, it was especially poignant that a keen African American mathematician like Katherine Johnson could share a telephone, or a slide rule, or a lunchtime game of bridge with her white male colleagues at Langley, while her daughters had to attend a segregated high school. On the one hand, she could openly discuss the *Brown v. Board of Education* Supreme Court decision with white colleagues, who tended to agree: It was high time to integrate the schools. On the other hand, in 1958, Virginia's governor chained the doors of any school seeking to obey the Court's integration order.[15]

Many Langley women faced similar dichotomies. NASA would eventually send Cathy Osgood to Cape Canaveral in preparation for launches—a woman traveling on her own for business. But she had to use her husband's credit card to book tickets, as wives were not allowed to have their own. As her NASA career bloomed in the years that followed, she found her nerdy husband to be a new kind of American man. "He was great. He's never one to say, 'That's a woman's job,' or anything of that sort." She recalled intense deadlines for orbital computations—"I did bring in my sleeping bag and sleep in the ladies' room"—where her husband would take care of the kids, groceries, and the cooking. (To be sure, Mr. Osgood's attitude fell well outside norms at Langley and in America. But the outlook was making its debut.)

In the end, nothing was more groundbreaking than NASA's absurd-sounding aims. In 1960, many Americans enjoyed dreaming of space, but many others wondered why the government would start spending all this money on such an esoteric project, Sputnik or no. "A man in space," a former boss said to Osgood, after hearing of NASA's first goal. "Doesn't that give you the creeps?"[16]

The one-time submariner Max Faget had taken on the role of spacecraft designer for the nation's first phase of space exploration, dubbed "Mercury," aiming to put an American in orbit around Earth, hopefully before the Russians. By 1960, he'd finished the basic design and it was full of surprises. This early spaceship shared much with a submarine—not only was it absurdly cramped, with no corner wasted, but it also had a periscope instead of a window. Faget and his colleagues even argued over the size and weight of the human pilots. Given the available rockets' ongoing struggles to lift much weight, it seemed absurd to send up, say, a six-foot, 180-pound test pilot when a much smaller person would do just as well. One engineer, a veteran who'd survived bombing missions over Germany, deadpanned that they should find legless astronauts to save forty percent on body weight.[17]

Faget seemed to have solved several basic problems for keeping a live astronaut rocketing into space and back to Earth in one unbroken piece, though the proof would await a real, nail-biting test. Not only would the tiny new capsule, sitting atop a rocket, have to protect a human body from harm in the violent acceleration toward space, it would later have to protect the human from an arguably more violent return to Earth, as the capsule would somehow slow from thousands of miles per hour, not burn up from the scalding friction of our atmosphere, and find a survivable impact speed.

The intense heating of re-entry was not a new problem. Early ballistic missile tests had measured tips of returning missiles surpassing ten thousand degrees Fahrenheit, well above the temperature of the sun's surface. Even von Braun's V-2 program had grappled with this issue. He and his team had considered all sorts of ways to cool a missile as it re-entered the thickest part of the atmosphere. He had even entertained a system with forced coolant near the nose cone, but introducing extra pumps, with fuel for those pumps, simply added too much extra weight.[18]

The surprising key to NASA's safe re-entry was having a dumpy, blunt shaped craft instead of a sleek, pointy one.[i] Returning from space, a more pointy craft will zoom into Earth's atmosphere and penetrate more deeply before slowing appreciably. Moving that quickly through the thicker air leads to the hotter-than-the-sun temperatures. Meanwhile, a craft shaped like a laboratory flask flying fat-side down slows more gradually. It also allows heat to dissipate into the surrounding air instead of building up in the vehicle itself.

To further outsource the incredible heat of re-entry, the engineers opted for a scary-sounding technology on the capsule's bottom: an "ablator," or a substance that literally burns away, sloughing off material as the spacecraft descends. As long as

---

[i] The original blunt-body idea is credited to Dr. Harvey Allen, working at the Ames Research Laboratory in the 1950s. Faget turned it into a viable space-faring module.

the capsule had a thick enough layer, the hot gas byproduct leaving the heat shield would actually carry a lot of the extra heat with it. And, as a small bonus, the capsule would literally become *lighter* as it re-entered; every ounce it could shed would make its journey just a tiny bit more gentle, moment to moment. But calculations and tests would have to be precise: too much ablator could make a capsule too heavy to launch, but too little meant the fires of atmospheric friction would consume the ablator and then roast the astronaut inside. "I did it on a slide rule," said Aleck Bond, a colleague of Faget's. "All my analytical work was done on a ten- to twelve-inch slide rule." The thermal analysis suggested that just one inch of ablator should work. But for early tests, Bond says they doubled it, just to make sure.[19]

By 1960, these tests looked promising. And, true to the slide rules, using just an inch-thick fiberglass laminate proved good enough. It would lose a few pounds of material during re-entry, but only a fraction of its thickness overall, while temperatures inside the capsule remained moderate. (For the eventual Apollo capsule, much larger than the early man-in-space capsules, engineers fashioned a steel honeycomb lattice on the capsule's bottom. Technicians then used caulking guns to inject each of 400,000 cells with a special new ablator that had been developed for nuclear warheads. This laborious attention to detail paid off; considering all of Apollo's myriad components, the heat shield was among the least troublesome.)[20]

Temperature was far from the only concern when considering human beings lofted to space and back. The incredible thrust during launch and braking during re-entry would challenge the structure of the human body. The efficient genius of Faget's capsule design had its fat bottom directed Earthward both during its launch atop a rocket and also for its solo return to the surface. At launch, the astronaut would be pushed against the capsule's fat-bottomed side, as the rocket underneath pushed him up and into space. (Consider your car accelerating very quickly and your seat back pushing against you.) Then, upon return, when the capsule needed to lose tens of thousands of miles per hour, Faget and his team had the capsule plummet backwards, fat end first, and the astronaut would again be thrown against his seat back. (In this case, imagine if, with your car moving at one hundred miles per hour, you could deftly spin it around backward, and as you slowed to a stop your body would feel the strongest pressure again in the padded seat instead of straining against the seatbelt straps.) For the incredible accelerations of a space mission, this method gave engineers the best odds for protecting an astronaut's ribcage and organs.

An automotive seat wouldn't do for a space voyage. Faget and his fellow engineers came up with the "survival couch," custom-molded to each astronaut's back, including the backs of arms and legs. Engineers recall early trials of the survival couch idea, using an unlucky set of farm animals. According to one account, a couple of the Yorkshire pigs died from their upside-down fall, and the engineers reworked the survival couch.[21] The redesigns smoothed the kinks and bumps from the seating, and watching healthy pigs trot away from horrific-looking backward falls raised the engineers' hopes. The pigs in their safety couches were now surviving accelerations many times what any astronaut would need to experience. Tests on humans followed. Test subjects could be accelerated to twelve and even twenty times the acceleration of free fall on Earth and walk away from the survival couch without injury. NASA molded a special version for each of the original seven astronauts, just like one might do for a mouthpiece or an orthotic shoe insert (see Figure 3.2).

The capsule's progress encouraged NASA, but their ability to pitch it safely into Earth orbit was another matter. In 1960, young Marlowe Cassetti became the chief reviewer for a series of disturbing silent films. "Somehow I was tagged as the launch guy," he says. "I would get on average one per week . . . they were shooting up a lot of rockets. . . . I would get an express pouch from either Florida or California, hand delivered to me. I think they may have been classified. Our probability of getting into orbit had to be classified too." A technician would lead Marlowe to a conference room, help set up a reel-to-reel projector, and then leave him with that week's film, usually a compilation that played like a grim blooper reel. "If I recall correctly, they were nearly all failures. . . . [The rocket] would go up a few feet, fall back down, and

FIGURE 3.2 Test versions of custom-molded survival couches at Langley. The names here include those of Langley employees who volunteered as test subjects. (NASA photograph.)

there would be a tremendous explosion." Marlowe quietly watched the debris settle before a rough splice brought on the next launch.

Reflecting on those early times, Henry Pohl summarized the situation: "The criteria for the launch to be successful was that it got out of sight before it blew up." And as von Braun preached to Henry and his other Huntsville troops circa 1960, the initial goal for their rockets was to "make the target area more dangerous than the *launch* area." Calling progress in rocket engines from 1958 to 1962 "phenomenal," Pohl says it's easy to forget just how bad things were. "In the late fifties, I believe one year we had fifteen Thors blow up. . . . They must have built about sixteen Titan I's, and I am not sure that any of those ever got away from the pad. Maybe one."[22]

What kind of notes did Marlowe Cassetti take during his grim private screenings? Mainly he kept track of successes versus failures and scribbled abbreviated observations for his bosses. *One success, three failures, one abort (with no explosion).* And Cassetti thought to himself, "We're going to put guys on top of this? It's really crazy."

But Faget and his team had a plan for the worst-case scenario as well. They designed an "escape tower." Photos of some of these rockets, including the eventual Apollo missions, show a skinny extension from the rocket's highest tip, as if the designer were just trying to provide another few feet of height. In the event that a rocket blew up, the ingenuous escape tower, perched atop the astronaut's capsule, would fire its own little cluster of rocket engines and *pull* the capsule away from the exploding rocket and deliver the astronaut to safety. Timing would be everything, but the engineers were confident the system could work. (As the program evolved, the tower was never needed, but by copying this system, the Soviet Union did later save the lives of two cosmonauts. They flew to Houston many years later to thank Max Faget in person.) (See Figure 3.3.)

Fundamental designs marched rapidly forward, thanks in part to a key set of engineering immigrants. Canadian aerospace engineers had first unveiled an advanced jet fighter, the Arrow, on the ill-fated day of October 4, 1957. "We rolled [it] out of the hanger for the press to see," recalled engineer Owen Maynard, "and hardly anybody showed up at the stands." Everyone covering aerospace for the media had been consumed by the news of Sputnik. That was the beginning of the end for the Arrow. After the Canadian government canceled the jet in 1959, NASA invited about twenty-five of the engineers to bring their expertise to Virginia. This was no simple move for the Canadians. "You have to understand," one of Maynard's colleagues said later. "There's considerable prejudice north of the border about coming south of the border."[23] But the interesting work, at this point, pulled them across. Maynard, a structural engineer, specialized in craft that moved at extreme speeds under great stress, and he had helped design escape systems for jet pilots. "We were very quickly accepted royally into the community at the Space Task Group," Maynard said.

FIGURE 3.3 Artist's rendition of the one-seater Mercury capsule with the escape tower attached (*top*). The pack strapped to the heat shield in this drawing (*bottom*) helped the capsule maneuver when in space, but like the escape tower, the pack was (usually) discarded before the capsule returned to Earth. (NASA image.)

Before America's television habit, the engineers embraced old-world evenings. "We took ballroom dancing lessons with them," Maynard said. Once a week, the class would retire to one of their homes. "We would continue with playing records and doing this different kind of ballroom dancing, including the Twist." Aside from the dancing, however, his family suffered what so many did in NASA's decade to come. His wife would ask him, "Why is it that you spend so much time at work when you've got these three wonderful kids and you love your kids so much?" She was right, Owen thought, and he adored his children, but, "I never could answer her that." A great many of the engineers have expressed a similar wistful regret.

In the summer of 1960, Maynard witnessed the first test toward lifting a man into space. It aimed to use an air force "Atlas" rocket, just like those Cassetti had been watching, and lift a test capsule, sans astronaut, into space. Given the amount of work Maynard had already put in, NASA rewarded him with a trip to Florida to watch the launch. Maynard told an interviewer later about the strong emotions of those around him. "This thing was getting ready to fly, and it was like the baby that's grown up and he's going off to college, maybe going off to war or something. It was that kind of emotional." One of the engineers cornered him. He confessed to Owen that he was worried sick about how the rocket coupled to the test capsule above it, saying the connection presented a definite weak point.

Despite heavy cloud cover in Florida, NASA went ahead with the launch, and the Atlas came apart about one minute after leaving the pad. Because of the weather, the team had no clue by eye or on film as to what happened. From that point forward, NASA set a rule against launching into overly cloudy skies.

After the rocket came apart, the project engineers scrambled about, sifting what little data they had. Bob Gilruth came striding out of the control room, found the off-duty Owen Maynard, and put his hands on Maynard's shoulders. "Go find that thing," Gilruth told him. "Find out what happened." Maynard and some colleagues rented a couple of fishing boats and found the rocket's debris field at sea. When a scuba team struggled to find a couple of key parts, Maynard took a deep breath and jumped in, free diving to a depth of thirty feet. His familiarity with the shape and design of the parts helped him locate the rogue pieces. They hauled everything back to an airplane hangar at the Cape and began a painstaking reconstruction, trying to diagnose the failure.

A subsequent meeting brought together engineers from the air force (the parents of the Atlas rocket) and representatives from various contractor companies who had built pieces of the Atlas and the capsule. During a coffee break, Maynard had an idea. He took out his slide rule and started scribbling mathematical notes on, yes, the back of an envelope. According to his calculations, the thin skin of the Atlas rocket could crumple and fail at the most stressful moments of the rocket's ascent toward space—for instance, when the rocket broke the sound barrier about a minute into its flight. He handed his results to his boss, the only other NASA representative at the meeting. After coffee, his boss stood up and thanked everyone for their presentations, but he announced that Maynard had solved the puzzle and then abruptly adjourned the meeting. His scribbles, computing strain on the rocket at "Max Q," the most violent acceleration, showed an alarming chance that the rocket's thin skin would just crumple like so much paper. The others in the meeting "were livid," according to Maynard, but the answer was undeniable.[24]

When engineers describe the Atlas as a metal balloon full of rocket fuel, that is no exaggeration. Without the pressure of fuel puffing its thin skin outward, a thumb's pressure easily indents the side of a preserved Atlas today. Why build such a flimsy rocket? It goes back to needing to lift the least possible weight. The lighter a rocket, the less energy it needs to climb skyward. And indeed, when the Atlas was full of fuel, the skin was rigid enough, or almost rigid enough, for its workload.

The best solution for the rocket's thin skin bordered on humiliating for the rocket's military designers: a metal girdle for Atlas. "I made the mistake of calling it a 'belly band'," Gilruth recalled. "This was very, very unpopular with the air force." But it worked.[25]

Approaching Thanksgiving of 1960, NASA prepared for another unmanned test. After the failed summer launch, NASA had suffered another beating in the press, and political pressure on the young agency redoubled. Now, with one of von Braun's rockets, NASA would try again to lift a capsule into space, to see if it would leak, to see if it could function, to see if its interior temperatures would support a fragile human being, and to see if it could return to Earth in one piece. As anxious NASA administrators, their political patrons, and scores of journalists tapped their fingers, the launch team waited through an unbroken month of hard Florida rain.

Around midnight on November 21, when the countdown finally reached zero, the excitement hit a crescendo. As the engines fired, one of von Braun's German colleagues fell to shouting in German over the main communications channel, drawing English rebukes from engineers monitoring crucial systems. But the tremendous roar suddenly hushed, and as clouds of exhaust drifted away from the rocket, everyone could see it still sitting there. It had lifted about four feet from the pad and decided it felt like sitting back down; the rocket apparently couldn't be bothered with a launch today.

As network television rolled, the little capsule, atop the now dormant rocket, sprung to life. It thought it must be landing already, so out popped its automatic parachutes, which draped downward over the rocket, as if protecting embarrassed modesty. And then, completing the sad display, a canister of green dye burst forth, intending to mark an aquatic landing spot to aid recovery teams. An engineer muttered something about the similarity to a clown car at the circus.[26]

While the engineers could only wish the malfunction had somehow stopped the TV cameras recording this feeble display, they had bigger problems. They had a rocket packed full of kerosene and liquid oxygen sitting upright with nowhere to go. And the wind might yank the entire tipsy thing over by its misguided parachutes. NASA did not know of the Soviets' Nedelin disaster, but the possibilities of an unhappy ending were clear. One voice asked to get a gun, suggesting they shoot holes in the tanks to drain them. In the end, the engineers decided to keep everyone away from the rocket and let the returning Florida dawn eventually boil off the liquid oxygen; with that half of a potential disaster removed, crews then snuck out to drain the kerosene.

Engineers cleared just enough of the egg from their collective face to focus on what went wrong. Within weeks, the team had busily diagnosed the sensitive rocket's premature shutdown and prepared another test. Just in time for Christmas, they had their first success, lifting one of Faget's empty capsules into space and having it safely parachute into the Atlantic Ocean. The rocket worked. The capsule endured its moments above the atmosphere without leaking air. The heat shield kept it from burning up on return. Finally, the capsule survived impact on the waves and floated long enough to share its story. NASA had wobbled to its feet.

4

1961—A TODDLER'S MARATHON

Max Faget recalled some of the earliest discussions of man-to-Moon missions, starting within months of the formation of NASA. On short notice, NASA headquarters asked him how *he* would recommend flying to the Moon. "So I spent a couple of hours before I went to Washington that time, thinking about it," he said.

A team of engineers from the Jet Propulsion Laboratory in Pasadena, California, also attended. They had already been considering Moon missions, planning to send unmanned probes (the "Surveyor" program) to land there and take measurements. They recommended their ballistic approach for any manned landings as well. "You know what the Surveyor does," Faget recalled. "It comes screaming in at the Moon at something like about nine thousand feet per second [over six thousand miles per hour], and when you get about twenty miles high, they turn on the rockets and slow down." Faget saw that as too risky, especially with humans on board: slow down too soon, and you'll waste all your fuel; slow down too late, and you'll slam the mission to pieces on impact.

He proposed a more cautious, step-wise approach. "I got up and said, 'Well, the first thing you'd do is you'd fly around the Moon,' and told them how much velocity that would take and all that, made the typical figure-eight drawing, you know. You'd just go out there, fly by, and you'd come on back, and we wouldn't go very close to the Moon. We didn't know exactly what the gravity number of the Moon was." He compared it to something more familiar: flying a new breed of airplane. "First you taxi it, then you take off, and you maybe go up to four or five thousand feet, you fly around the airport a couple of times."[1]

A debate emerged over the exact conveyance that would land humans on the Moon and bring them back. At first glance, most people assumed that we would build a sleek ship, launch it up into space using a three- or four-stage rocket, fly it to the Moon, carefully land the ship there, butt side down, later launch *again* (an easier launch, since the Moon's gravitational pull is just one sixth that of Earth), and finally fly home. Such visions spilled from science fiction magazines, with silvery, planet-hopping ships. Why complicate matters?

All these single-craft approaches were called "direct ascent" in the debates, but a single-ship trip to the Moon faced a number of practical hurdles. First, the rocket

42

would have to hold all its fuel for the trip to the Moon and back, including—and this is the real killer—all the fuel needed to lift itself from the Moon. Meanwhile, to start its journey with one of von Braun's new and evolving Saturn rockets, the entire ship would also have to be somewhat thin, relatively speaking, to fit on top of the rocket. To be appropriately skinny while also holding the crew of humans, all their supplies, and all the round-trip fuel, this spaceship would then need to be tall. That created another problem: a necessarily unsteady spaceship. We couldn't assume the Moon's surface would be perfectly flat, and a tall ship risked tipping over, forever trapping astronauts in a lunar casket. More fundamentally, however, the planned Saturn rocket series could never lift something heavy enough for a direct ascent plan in the first place. The Huntsville team began exploring an even more ambitious rocket, called Nova, but in 1961 it was nearly science fiction.

The main competition to the straight-shot approach suggested multiple rocket launches, each lifting a piece of the eventual Moon-bound ship, and then having these pieces link together as they orbited Earth. After coming together, the unified ship would then trek to the Moon and back. This was called "Earth orbit rendezvous" and introduced a new, risky maneuver that led many engineers to shake their heads: take two (or more) objects moving faster than any human-made object had ever moved—say ten or twenty *thousand* miles per hour as they ran laps around Earth—and have these objects knock into one another and link up without damaging the thin hulls, explosive tanks of fuel, or any of their fragile contents.

Myriad other technical complications followed. Objects could inhabit an infinite number of different elliptical and circular shapes to orbit a planet. How could we launch *two* different objects and get them to inhabit exactly the same orbit so that they could dock with one another? Also, to make such predictions, our ability to compute the arcs of these ships would have to dramatically improve. Our planet is not actually a perfect sphere—it suffers a middle-age bulge about its waist—and since Earth is not a completely smooth rock inside, its gravitational pull varies slightly from location to location. Real orbital paths are, in their minutiae, much more complicated than they would be around a perfectly spherical and uniform planet. Furthermore, Earth's atmosphere doesn't quite *end* at a certain height; it more fades away. So, any object making loops around Earth is still encountering a thin, wispy gaseous atmosphere that is slowing the object down. We'd need to perfectly understand that if we wanted to achieve rendezvous.[2]

Even with its intimidating levels of complexity, Earth orbit rendezvous looked more possible than direct ascent in early 1961. But a sort of stepchild of Moon plans lingered on the sidelines. "Lunar orbit rendezvous" also involved a modular approach, with multiple ships taking on different roles and equipment. In this case, a special moon vessel, exceedingly light and nimble, would depart from the main spaceship, descend on a "lunar excursion," and then pop all or part of itself back up-ward. Once it rejoined and attached itself to the main ship, they would travel back to Earth. A light little craft descending to the Moon featured a certain logic because it took much less fuel to launch itself again. But the idea was the craziest yet. "No one wanted to do rendezvous at the Moon," Faget said, "simply on the basis that that's a pretty ticklish maneuver." If the engineers hadn't yet figured out how to track a spaceship and reliably communicate with it as it orbited *Earth*, why would they try a complicated maneuver with *two* ships that were a hundred times farther away? Thinking through these little craft in lunar orbit, engineers realized that any ships orbiting the Moon would spend substantial time blocked by the Moon itself, com-pletely cut off from any radio link to Earth.[i]

Through the spring of 1961, these thoughts of a Moon mission were speculative, per-haps decades away. The former president had expressed reluctance over a big-money space program. The newer administration hadn't shown much excitement for space either, but it did strike a different, more futuristic tone. As President Eisenhower left office, he granted an interview, wherein he worried over the growing clout and intermingling of technology and defense. "When you see almost every one of your magazines, no matter what they are advertising, has a picture of the Titan missile or the Atlas," he said, "there is . . . almost an insidious penetration of our own minds."[3]

The incoming White House politely viewed that thinking as outmoded. John F. Kennedy and his vice president, Lyndon Johnson, largely ignored Eisenhower's warnings of industrial technology and large, government-funded scientific and mil-itary projects. Many historians have put Kennedy and Johnson among the first and most influential disciples of a "technocracy"—better living, better governance, better *everything*, through a systematic embrace of technology.

Despite a bright-eyed new administration, job candidates did not line up in 1961 to direct what was seen as a struggling new space agency. Johnson later recalled sev-enteen different individuals saying "no thanks" to running NASA. And after he sup-posedly twisted the arm of one James Webb to take the reins, Kennedy informed the new director to expect no new funding for that Moon mission idea. In early 1961, Kennedy ranked the fledgling "Apollo" project—named just months earlier for the chariot-riding God of prophecy, light, and progress—as a fanciful solution in search of a problem.[4]

More practically, NASA made progress with near-Earth plans. They'd successfully, if traumatically, kept a chimpanzee named Ham alive during a shot into space. Once

---

[i] These were hardly the only options. One serious contender involved landing two ships: a light one with astronauts and another one nearby acting as a pack mule with fuel and supplies.

aloft, Ham had work to do, as NASA doctors sought to understand how well his mind worked in orbit. He was to get food rewards for completing tasks correctly and little electric shocks to his feet for wrong answers. Unfortunately, the experiments went haywire in orbit, regularly zapping poor Ham no matter what he did. While Ham's spaceflight and return to Earth were awful enough, his recovery at sea might have been the most violent. He hit the water so hard that his capsule sprang a leak. And of their many problems to solve, NASA had underrated the difficulty of snatching a floating capsule from the rolling sea. "When you pick [the capsule] up," said recovery engineer Peter Armitage, "it becomes essentially a wrecker's ball, and it bangs into the side of the ship. Poor Ham got more G's [strong accelerations] from the recovery than the whole space flight."[5]

But the program inched forward. An Atlas missile, with its humbling new belly band, stayed in one piece and carried another unmanned version of Faget's one-seater capsule into space and back. By springtime, they were ready to risk the real thing: lofting a human on the same, brief arc.

On April 12, another shock arrived: The USSR commandeered the world's headlines again and beat NASA to another punch. The Soviet Union launched a small Vostok[ii] craft into a single orbit of Earth. Yuri Gagarin became the first human being in space and the first to orbit Earth. Soviet engineers had put the entire mission on autopilot. If Gagarin had lost consciousness, or in case he'd entered some sort of delirium, they didn't want him to screw things up. (Rocket designer Korolev felt for Gagarin, however. He'd created a sort of manual override option and slipped a 3-digit "unlock" code into a sealed envelope in case Gagarin needed to adjust controls manually.) Though photographs weren't released for days, the Soviets announced that their mission had returned Gagarin safely to Earth. American engineers debated how their rivals were able to set a heavy spaceship directly on land, but in details that only emerged years later, Gagarin and other early cosmonauts ejected from their spacecraft and parachuted to the ground while their ships made violent impacts with Earth.

Reporters called NASA's PR representative at 4:00 a.m. local time in Virginia seeking a response. "We're all asleep down here," he accurately said, but the negative headlines then wrote themselves.[6]

Later that morning, a young President Kennedy, still getting his feet set in a most difficult job, paced his White House office. Unbeknownst to most Americans, he faced debilitating health problems, including a long fight against an adrenal illness. (Not

---

[ii] This word, interpreted as "east" by the Western press at the time, can also mean "an upward flow" and in some sense suggests a sunrise.

long after Kennedy's post-war tour of Germany, a British physician had worried that the young American might not live another year.) According to his wife's complaints in 1961, he was also obtaining under-the-table Demerol—a narcotic painkiller with known psychological side effects—from a secret service agent. He stood and composed a rapid-fire, multi-point memo full of questions for Vice President Johnson. His first item: "Do we have a chance of beating the Soviets by putting a laboratory in space, or by a trip around the moon, or by a rocket to land on the moon, or by a rocket to go to the moon and back with a man? Is there any other space program which promises dramatic result in which we could win?" Another item in this memo was also telling for the future of NASA employees and those of hundreds of contracting companies. "Are we working twenty-four hours a day on existing programs? If not, why not?" Johnson, over the next couple of weeks, composed a reply. He spoke with a number of people, including Wernher von Braun and, interestingly, the president of media company CBS.[7]

But before Johnson could deliver his results to Kennedy, another PR disaster greeted America the following week. Kennedy okayed the infamous Bay of Pigs invasion of Cuba, which failed as completely as it did quickly. Engineers recalled working to set up a communications station in Guaymas, Mexico—one of about a dozen stations in the works around the globe, so that NASA could stay in regular contact with orbiting space ships. They had just finished an embassy briefing that spring on how to build a working rapport with local government officials. First, the engineers heard of the Russians putting a man in space, and they worried the locals might think NASA was second-rate. A week later, America had invaded a nearby Spanish-speaking neighbor, while the engineers awkwardly cleared their throats and continued to negotiate an allegedly peaceful communications station.[8]

Johnson's recommendations on space arrived on Kennedy's desk a week later. They included a dramatic increase in funding for NASA, and he noted that, in particular, a Moon landing would have "great propaganda value." While we can convincingly connect a path from von Braun's advocacy to Johnson's spring 1961 statement that "second in space is second in everything," and from there to Kennedy's Moon decision, other channels were active as well.[iii] Bob Gilruth, the leader of the Space Task Group, recalled direct talks with President Kennedy in spring of 1961. "Kennedy came along and said, 'Look, I want to be first. Now do something.' I said, 'Well, you've got to pick a job that's so difficult, that it's new, that they'll [the Soviets] have to start from scratch. . . . Going to the Moon will take new rockets, new technology, and if you want to do that, I think our country could probably win.'" Both von Braun and

---

[iii] The broad brushstrokes presented here do not fully relay a complex decision, analyzed by book-length manuscripts over the years.

Gilruth had major influences, by all appearances, but von Braun may have had more impact on the eventual timeline. For years, he had produced a steady stream of ambitious goals with date stamps, and his reply to Johnson emphasized the possibility of getting to the Moon before the year 1970.[9]

A week later, a NASA test flight went awry. An Atlas rocket held together, but this time it shot *too* straight. Engineers typically have rockets launching from Florida roll and turn as they rise, flying out over the Atlantic on their way to space. (This practice moves a mission toward an orbital path, and it also avoids a failing rocket taking out hundreds of unsuspecting Floridians.) But on April 25, the rocket's internal guidance system failed and it never altered its course. After watching the rocket go straight up above Florida for about forty seconds, NASA's control team sent a destruct signal so that it wouldn't hurt anyone; the mission became an expensive firework. As a silver lining, Faget's escape tower worked perfectly, rocketing an empty capsule away from the Atlas explosion. The capsule then drifted by parachute to the sea, where NASA recovered it for use in later missions.

Cold War paranoia infected the aftermath of every misfire. A mysterious Russian "fishing trawler" had started attending launches, floating near Florida. Some engineers openly wondered if the boat, bristling with antennae, might be sending disruptive signals, but NASA never found evidence of foul play.

Meanwhile, engineers fine-tuned a rocket for, at last, launching an American into space—not a chimp but a human being. Cassetti, Pohl, and their colleagues did their best to push all the earlier rocket explosions from their minds.

On May 5, NASA learned that while the normal parade of countdown delays was not necessarily a problem for machines, it might take its toll on a human being sitting within a tiny capsule, seventy feet above the ground, with everyone watching and waiting. Some spectators of that launch said they became increasingly concerned about the poor, brave astronaut, Alan Shepard. After a number of hours, our first astronaut finally, with no complaints, urinated in his spacesuit. Despite thinking of every possible obstacle, the engineers, in planning what would be a very short flight, hadn't thought of a rocket-ready bedpan. Finally, the countdown made it past every delay. The rocket from Huntsville lit and held together. The mission's fifteen-minute parabolic arc into space took Shepard about one hundred miles east and safely plopped him and the capsule in the Atlantic Ocean as a wholly new type of American hero. The nation cheered its first astronaut.

The good news encouraged President Kennedy and his inner sanctum. In preparing a major speech for a joint session of Congress on May 25, they decided to include a space-related announcement as a ninth and final bullet point. While we now remember this speech as promising to land a man on the Moon and bring him back to Earth by the end of the decade, that was appended to fifty minutes detailing progress and challenges, continent by earthly continent. The Apollo goal, as announced to the world in 1961, sat buried in the context of battling the spread of communism across the globe.[10] As he committed America to the Moon, Kennedy also committed to the "Alliance for Progress," sending money and agents into South America in hopes of short-circuiting leftist uprisings, and he spoke of increased activity against communists in Vietnam.[11] If the full-court, anti-communist agenda stretched across the globe, why not extend it beyond the atmosphere? For Kennedy and Johnson, the space program would establish a statement of superiority and proof of principle for peoples of the world who teetered on the fence, in terms of which superpower and which system they would follow.

We might assume today that the NASA engineers would have reacted to the public Moon pronouncement with cheers, and some, like von Braun and his inner circle, were indeed very happy. No city was more excited by the events of early 1961 than Huntsville, Alabama. At a celebration of the first American astronaut in space, von Braun quieted a boisterous crowd in Rocket City's Courthouse Square for an impromptu speech. "We will go farther and farther," he promised them, "eventually landing on the Moon."[12]

But in the engineering trenches, reactions to Kennedy's speech featured more variety. "I thought that was the dumbest thing I'd ever heard in my life," says Henry Pohl. "I mean, you have to appreciate where we were back in that day and time. . . . [W]e still had vacuum-tube technology." Pohl notes that most launches of America's largest rocket were still ending in failure. And longtime Faget collaborator Caldwell Johnson summarized the abrupt shift as follows: "It's one thing to sit around the table at noontime and play pinochle and bullshit. It's another thing for the president of the United States to all of a sudden tell the world what you're bullshitting about!" Many engineers noted the vast gap between what they had done so far (a brief *lob* of a man brushing against space) and what the nation said they now must do (a round trip of nearly five hundred thousand deadly miles). "It was just absolutely—we were incredulous," Marlowe Cassetti said. He laughed at the memory. "It's like, you know, I can think of a toddler taking a first couple of steps and being unsure, and then you say, 'In a couple of years, he's going to run a marathon.'"[13]

To get there, no matter which mode won out—direct or modular, with rendezvous at Earth or at the Moon—NASA knew they needed new rocket engines. Even the lightest possible versions of a Moon mission would need engines more powerful than those Korolev was building for the Soviets. The design and perfection of such monsters fell to Huntsville's Marshall Space Flight Center.

In fact, they were already working on such an engine, one that could, by itself, create a force equivalent to one-and-a-half *million* pounds (compared to about eighty thousand pounds from their earlier rocket engines). Such a thrust could hold aloft a large, three-mast clipper ship from the nineteenth century. Clustering a few of these engines in concert could, with luck, start a trip to the Moon. In the weeks before Kennedy's challenge, the rocket team had surprisingly good news. A prototype of their new engine had survived a short test and marked a record thrust. They had a long way to go—further than they even knew—but they had a start on what would eventually become the giant F-1 engine. Given the short deadline (years instead of decades), the engineers decided to use existing principles, souping up the basics of von Braun's V-2.

As Henry Pohl tells it, "One pound of [kerosene] fuel can lift 250 pounds for one second." So, the math and the challenge follow. The new engine would thirst for torrents of kerosene and oxygen brought together at unprecedented speeds and volumes. There would be no normal sense of "flow" to it, in the end. Enormous pumps would force-feed fuel to the engine's inferno.

Pohl recalls the first time he entered von Braun's main conference room. "He had all of his people . . . in some kind of pecking order," he says now. "But he had brass plates with their name on it with their place to sit. So each time they were sitting in the same seat." Other engineers have described the shifting fog of cigarette smoke in the room. Their chief ran efficient meetings, projecting calm, charm, and respect, smiling through tense moments of debate. Contemporaries, in recalling von Braun, often note his listening skills. He sought each and every opinion from the assembled and welcomed questions. His co-workers and deputies felt they'd had a fair shake.

Pohl recalls the first time he saw von Braun speak. At a conference in the late 1950s, von Braun lectured on one of his passions: a permanent, rotating space station. "When he got through," Pohl says, "he wanted to know if there were any questions, and I said 'Yes, you can't balance it.'" Henry had mounted his share of tractor and truck tires on the farm, and he knew it would be impossible to stabilize something with people and equipment moving around in it.

"Ah!" von Braun said. He rifled through his briefcase, found an extra slide and dropped it into his projector. He explained to Pohl that expertly placed water tanks would maintain the station's balance.

The structure of von Braun's organization reflected the structure of his team back in Germany, with separate laboratories exploring different aspects of rocket design, testing, or control. On his frequent visits, Marlowe Cassetti noted their incredible independence from one another, toiling away like separate fiefdoms. The laboratories, he says, "almost operated as standalone space centers."[14] The official history of the Moon rocket's development, the NASA-produced book *Stages to Saturn*, notes that the center's management style lacked a defining character but was "more of an amalgam of various concepts."[15] While the different laboratories had their own structure and style, all would agree that they shared a central principle: Managers, directors, and team leaders owned engineering skills themselves. They could roll up their sleeves, don lab coats, and place their hands on the equipment to probe and troubleshoot any technical detail.

No matter their independence, von Braun liked to visit the different laboratories, often unannounced. One of Henry Pohl's first encounters with him started with a misunderstanding. "It was just about sundown, and this old '51 OD Chevrolet . . . come up from the piney woods. We had a gravel road that came in from the Tennessee River." Personal cars weren't allowed anywhere close to the test lab. Henry went out with a megaphone in one hand, waving the other. "Get that thing out of here. . . . Get it out!" Pohl yelled. "He just turned, come driving right up to me. He got within ten feet of me before I realized who it was, and, of course, then I got tongue-tied." But cordial as always, the center director wasn't annoyed; he just asked Henry to show him his current work. Pohl took von Braun into the laboratory, where he was using a set of tiny thermometers to follow the way heat flowed through an engine system. Pohl described every detail, including new pieces he'd designed just for this task, and von Braun absorbed it all. "He said, 'Very ingenious, ingenious,' and went on off," Henry says. "Got in his car and drove off somewhere else."[16] This pop-in approach wasn't always welcome, and when it came to working with the many contractors who would build the pieces of the Moon rocket and its engines, von Braun's center was known for invasive oversight and laying a heavy hand on a private company's internal workings.

For someone in charge of building the world's most powerful rocket system, von Braun could also be surprisingly flaky and even petulant around non-rocket technology. One coworker recalled fixing the color balance on a television in von Braun's office. "What did you do? What did you do?" von Braun exclaimed, marveling at having actual colors instead of just sepia tones. "It's been that way for five years!" Later in his life, von Braun would literally tear VCR machines apart when he couldn't get them to function properly.

Overall, stories of von Braun losing his temper are rare. One arose when two of his colleagues continued sniping at each other, trying to score points and prove one another foolish. He barked at them to cease and desist. He closed the meeting and ordered his deputies to reconcile their views before they all next met. Another story saw von Braun anxiously awaiting a key film from orbit; he needed the footage to promote the work of his Marshall Space Flight Center. Scientist Tom Parnell received and cued up the reel for von Braun. But a problem struck the two of them as soon as they started the projector. The other NASA center, a newer outfit near Houston, Texas, had mounted the entire film backward, rendering it useless for von Braun's purposes. Sharp invectives followed. "That's the only time I saw him pissed off," Parnell says.

But the Houston people didn't do this on purpose, did they? "I thought it was quite likely," Parnell says. "There was an intense rivalry between the centers."[17]

In 1961, NASA christened a new Manned Spaceflight Center, to be directed by Langley's Bob Gilruth. Its aims were to coordinate the manned space missions, including the construction and testing of the spacecraft, the planning of the flights, including lunar explorations, and the training of the astronauts. Placing the new Manned Spaceflight Center for Gilruth and company involved a necessarily political process. Marlowe Cassetti recalls rumors of various sites, as they huddled over a U.S. map in their Virginia office.

The site selection committee followed a supposedly clear process. According to Marlowe, a winning site needed: ocean or deep-river access, because some large rocket and space craft pieces might need to be moved by ship; proximity to at least one major university; access to a substantial local workforce; and, among other things, year-round easy flights for the relatively new commercial airlines business. (Ease of travel increased the odds for a southern or western site.) Within a whirlwind of two weeks, a selection committee visited twenty-three sites in all. The top choice, after deliberations, was Tampa, Florida, where the air force had planned to close their MacDill base, making a large, developed site readily available. At the last moment, approaching mid-September, the air force opted to keep MacDill open, effectively removing Tampa from the list. The runner-up site became the top choice.[18]

On September 19, NASA announced the result: a swampy cattle pasture south of Houston, Texas. Here, they would build a new campus, the Manned Spacecraft Center (eventually renamed the Lyndon B. Johnson Space Center). As designed, the modern, multi-building facility could transition into a college campus after the possibly temporary space program came to a close.

Many people have long assumed that the center is only in this location because of Vice President Johnson's preference, but a more likely influence came from Albert Thomas, the chair of the House Appropriations Subcommittee, who happened to be a long-serving representative from Houston. Most accounts include local Rice University simply "donating" a large swath of pasture to the federal government for the new center. But the reality saw vintage Texas real estate in action. The Humble Oil Company donated the land's surface to Rice University, while maintaining their rights to what they needed, the oil and gas below. The company enjoyed a big tax break for their generosity, and Rice pivoted to make a triple play: They kept some of the land, donated some of it to NASA, and sold another chunk to NASA for $650,000.[19]

This flood-prone area sitting between the small towns of Webster and Kemah, some of the flattest and most humid land on Earth (with a sticky, black "soil" the locals call "gumbo"), borders an opaque, dishwater body of water called Clear Lake.

I share this homeland with thousands of other children from the Apollo era because of the 1961 federal decision. NASA's new site was a segment of bayou country still rooted in rice farms, cattle, and shrimp boats. The land oozed various snakes and every size of insect a child could want. My earliest memories include hot nights on a screened porch, with the keen of cicadas so loud I wanted to cover my ears. It was a great climate for spending most childhood days outdoors. Kids could chase crawdads in the perpetually full drainage ditches and go fishing in all sorts of little ponds and creeks. But I also remember the droning of distant, growing freeways. NASA aside, Houston's suburbs eventually ebbed to our doorsteps and overtook the area like a storm surge.

We knew it wasn't *beautiful*, exactly, but even as an adult I was a little surprised to hear the reactions of the Space Task Group. Some of them moved from Virginia as soon as the site was announced. New York native Marlowe Cassetti will never forget it. "It was such a dismal looking place," he says. "And I thought to myself this is the end of the world. It almost moved me to tears." Marlowe's colleague Hal Beck recalls coming from "gorgeous" Virginia and checking the new digs in 1961; he "drove down I-45 and took this dirt road over to the site. It was gloomy, dreary, ugly and all you could see around you were—nothing; shrubs, and no pretty trees, no anything."[20]

With perfect timing, concurrent with NASA's announcement, hurricane Carla churned through Texas in September of 1961, devastating many of the low-lying fishing communities surrounding Clear Lake. When Bob Gilruth and a Langley team visited the area one week after NASA's announcement, they found debris everywhere. Pieces of structures littered the main pasture, and Carla had snapped some of the sparse trees in half. Engineer Aleck Bond recalled looking for houses with his wife, just a few months after Carla. "All you could see was bare slabs of homes that had been swept away by the hurricane. That was kind of disturbing. . . . We went back to Virginia, shaking our heads, wondering what are we really getting into going to a place like that. But we made the move."[21]

The area surrounding NASA's new site had little housing available. Some of the Langley group chose to live in a tiny Quaker community called Friendswood, because, while you couldn't buy cigarettes in the local grocery store, at least they had trees.

While bulldozing and surveying started in the cow pasture, NASA hurriedly rented clusters of office space around Houston. "What a mess," Faget's design partner Caldwell Johnson recalled. "It was a building over here and a building over there. . . . One of them even was an apartment building, and they knocked all the doors off the different apartments so you could kind of walk between one and the other, you know, and sometimes your office would be in the kitchen. I mean, literally, we'd be in the kitchen." This ramshackle reality couldn't square with the high-tech gleam most Americans assumed at the time.[22]

Given the scope of the new center, the largely unspoken tensions multiplied between Gilruth's enterprise and von Braun's outfit in Huntsville. The rocketeers were now confined to just the first few minutes of any eventual Apollo mission: the launches. The new center, and Gilruth, would take over from there. Engineer and eventual center director Chris Kraft had nearly come to blows with von Braun in their first meeting, as they argued about the design of Mission Control. He wrote that von Braun "always seemed rankled that he couldn't run the whole show." But tensions ran more deeply than that.

"It seems strange to be working with the same people we hated during the war," Kraft said to Gilruth over lunch one day. He asked Gilruth what von Braun was like. "Von Braun doesn't care what flag he fights for," Gilruth replied.[23] The simmering conflict grew in 1961, with von Braun sensing a new bright line between Huntsville's workhorse rocket on the bottom, and Houston's more glamorous space capsule on top.

Despite the gritted teeth of the leaders, old enemies often fraternized well in the trenches. Henry Pohl speaks fondly of his old German bosses and coworkers in Huntsville. He and his wife named their son Karl, for instance, after imported engineer Karl Heimburg.

Pohl tells a story about working with a contractor on the engines. He and this fellow would sometimes go waterskiing on weekends, and Henry noted "all these little scars, about that big," holding his fingers apart two inches, "all over his legs." Henry inquired, and the man said he'd been a gunner in an American bomber over Germany. Because the airplanes sometimes grew stuffy, he and his flight mates wore shorts. But in battle, trying to ward off German fighters, the gunner would find himself "up to your waist in brass," the red-hot metal casings ejected from the machine guns. When Henry's German boss, Guenther Haukohl, learned of the man's past, he asked for more detail. The American relayed his wartime whereabouts and missions.

"Ja, ja," said Haukohl. "There was a full moon that night, clouds were at seven thousand feet. . . and those bombers were just on top of the clouds."

"We never saw what hit us!" the American said.

"Ja, ja, ja!" said Haukohl (according to Pohl's retelling, complete with affected German accent). The German had been flying an experimental *jet*-powered plane, attacking the American bombers at bewildering speed. The plane could only run for about twelve minutes at a time, but the pilot claimed he'd had three loads of fuel and downed three bombers that night.[24] There seemed to be no hard feelings between the two, a scarred gunner and a blunt German who had once nearly killed him, as they now worked elbow to elbow on spaceflight.

In addition to Huntsville and Houston, another NASA effort bustled in the center of Florida's Atlantic-facing coastline. The agency scrambled to build a Moon-worthy launch facility among the orange groves of Merritt Island, adjacent to the Cape Canaveral Air Force Station, site of various rocket launches since 1950. The military had decided to move some rockets there in 1947, after a Wernher von Braun mishap in the desert. From the White Sands launch facility in New Mexico, one of his V-2–inspired rockets took a sickening wrong turn and left a deep crater, fifty feet across, near Juarez, Mexico. The Joint Chiefs determined that the Atlantic Ocean would provide a wider and more forgiving studio from which von Braun could paint the skies overhead.[25]

Writer Tom Wolfe described the Florida "soil so sandy that the scrub pines had trouble growing fifteen feet high, and yet malarial and so marshy that the cotton-mouth moccasins stood their ground and stared you down,[iv] the sort of hopeless stone boondocks spit where the vertebrates give up" to the insects. "At night some sort of prehistoric chiggers . . . rose up from the sand and the palmetto grass and went for the ankles with a bite more vicious than a mink's." Of all places, this was to be the starting point for mankind's loftiest ambitions.[26]

Engineers working near the Cape in the late 1950s still remark on the wildlife. Don Woodruff, charged with all things electrical for rocket launches there, said that when running or checking new electrical lines, to and from the rocket stands and control bunkers, "you made sure it was a cable you were grabbing and not a rattle-snake." He described one dramatically ill-fated launch, with an Atlas going in a sick-ening, unplanned loop overhead, and slamming into the earth with an explosion not far from their block house. Woodruff and his colleagues went out to pick through the wreckage, littered with roasted snakes.[27]

Some said that working at the Cape in this era felt like living through a car wreck. In preparing for all the launches of the man-in-space program (Mercury), engineers worked absurd hours. One recorded an average of nineteen hours and seventeen minutes per day. NASA administrators had to order a cap of twelve hours per shift, in order to fight off exhaustion.[28]

With an accelerating space race, there was no time to waste on Merritt Island. They would need ridiculous facilities, on the scale of the Great Pyramids, rising from soft coastal marshes and scrubland. They would need an assembly space, where pieces and stages of a Saturn rocket, along with a Moon-ready spacecraft, could gin-gerly come together, out of the weather—a garage, in essence, that could easily house the Statue of Liberty. For the fires of a Saturn launch, they would need a solid metal

---

[iv] Referring to *agkistrodon piscivorus*, a nightmarish, semi-aquatic pit viper living throughout the southeastern United States and especially near NASA centers.

flame deflector, shaped like a plow blade, forty feet high. Rising above it, they required a launch tower about four hundred feet tall, with sturdy, retractable arms that would release the rockets to space.

Given ambitions of a biblical scope, engineers weren't the only overworked employees. Two unions staged walkouts in late 1960. Wernher von Braun met directly with the union members one November evening. He emphasized helping NASA maintain its commitment to the nation, but found himself shouted down. A strike of 650 electricians, plumbers, and carpenters followed around Thanksgiving. NASA retreated, agreed to mediation, and hired a labor counselor.[29]

Aside from the structures, engineers also had to figure out how to transport an assembled rocket from its gigantic garage to a launch pad. Why would they try to move an enormous, expensive, teetering rocket—350 feet high but only about thirty feet wide? "They designed the Cape for thirty or forty launches a year," my father says. This "mobile launch" concept, championed by Wernher von Braun, guided facilities planning. While a team of engineers would someday be launching one rocket, another team could be safely preparing another in an assembly building a few miles away. Instead of just assembling rockets on the pad and shooting them off one at a time, a mobile system allowed a much faster turnaround time. Hypothetically, even a major launch disaster could be swept aside while a new rocket and its own tower, undamaged, rolled out a week later. In fact, von Braun envisioned launching a major Saturn rocket every two weeks. The Moon was always just stop number one for von Braun. He assumed a major space station would follow, and then a trip to Mars. Once we started, logic and momentum would never let us stop.

In 1961, NASA didn't have many compelling ideas of how to actually *move* one of these rockets, bolted together and upright, several miles toward a launching pad. Would they build a special rail line, use some sort of fat barge on a canal, or what? Even if they moved one of these rockets without fuel, it weighed about two hundred tons (the weight of twenty stacked elephants, or two adult blue whales balanced upright). Engineers calculated that it would fracture the steel wheels of a rail car. Even just moving the individual rocket stages to Florida left engineers shaking their heads. One Huntsville team seriously considered using dirigibles to move chunks of the rocket across the country to the Cape.[30]

A very different sort of conundrum, spanning the globe, saw major progress in 1961. At the inception of NASA, engineers knew that a launch from Florida, heading east, would lose radio contact with Cape Canaveral in a matter of minutes, and an orbiting astronaut, even with a carefully chosen loop around Earth, would only fly over American soil for a tiny fraction of each ninety-minute orbit. At the time of NASA's birth, however, no global communication system existed. There was nothing, from the military to the world of global finance, on which engineers could easily piggyback a network that maintained regular contact with an orbiting satellite or spaceship. Moreover, existing radar systems, bouncing radio waves from an object to locate it and see where it might be going, were designed to find missiles and jet aircraft, not objects zipping around in space. And finally, there was no way to reliably move ever-larger clumps of information around the nation and world. NASA needed to collect information from their spacecraft and input it quickly to their fledgling computers. There, the machines churned through future projections of the spacecraft. Otherwise, unwelcome surprises could escalate to tragedies within a single lap around the globe.

When forming the agency, Congress set aside a former agricultural research center for a "space projects center," and this grew over time to the Goddard Space Flight Center,[v] handling tracking, data streams, and communications for the manned missions.[31] Within a few years, one in nine NASA employees worked on these communication and network issues, with some deployed overseas. Handshakes were made with local governments from Australia to Nigeria. For a tracking station in Guaymas, the agreement between the United States and the Mexican government marked a true milestone, as one of the first of its kind between the two neighbors since the early years of the twentieth century.

By the summer of 1961, NASA had formally opened a worldwide network of communication stations. Now a satellite, or a man in space, would have regular (though not constant) contact with engineers on the ground; NASA could track an orbiting spacecraft, send it direct radio orders, receive sensor data from it, and carry on a conversation with an astronaut. Even then, some of the first eighteen stations in the first network were just ships floating at sea. To monitor the health of a person trying to survive in orbit, doctors had recommended a network that provided *constant* contact with any manned spacecraft, but such a system was neither geographically nor politically feasible. NASA settled on the rule of having no more than ten minutes at any time without direct contact, and these gaps would still provide some nail-biting episodes in the early missions.

We can appreciate the gulf between present and past via the stories of an engineer who helped bring the system to life. "You couldn't talk to the flight director at the Cape from [half] of those [global] sites," Arnold Aldridge said. "They only communicated via teletype. So at these sites. . . you would write out what you wanted to say on a piece of paper, you would hand it to a teletype operator, and he'd run

---

[v] Of all NASA's centers, Goddard developed the most heavy-metal-sounding acronyms, including GRARR (Goddard Range and Range Rate) and SATAN (the Satellite Automatic Tracking Antennas).

over to the teletype machine." A similar set-up at Cape Canaveral would receive the messages and run them to the right person before crafting a teletype reply. "Now, that whole process might take somewhere between three and five minutes if it was working well," he said. Given that a quickly orbiting capsule might only be over a station for seven minutes, some of the stations would only get one shot at dialogue with Mission Control in Florida—not yet in Houston—before sending a single message to the mission above. Worse still, no technology existed to transfer and gather exact radar tracking data. From remote sites, data tapes came to Goddard via couriers. Precise knowledge of orbital paths for the early missions wouldn't come together until a month after the mission's conclusion.[32]

Young Marlowe Cassetti, a phone line, and a pencil formed one early communication relay. For some of the early, unmanned test flights, Cassetti helped compute the exact timing for a capsule to leave orbit and head for a properly located splashdown. Someone at NASA's main data center in Maryland would read a mission's latest whereabouts and its speed over the phone to Marlowe (in octal, versus decimal, numbers). Cassetti wrote these as quickly and legibly as possible and handed them to a secretary waiting just over his shoulder. She in turn ran those to a key-punch machine to create a computer card that she rushed by hand to a computer operator. The computer then, if functioning properly and if given a flawless card, could compute the exact times for a capsule to fire its thrusters and come home.

While the process sounds error prone and slow, Cassetti says that, after frequent interruptions of "wait, read that again," it worked well. "Interestingly enough, we could compute retrofire time as fast as they were doing it at Mission Control in Florida."[33]

The summer of 1961 also witnessed the completion of NASA's most critical early hard-wired data link, from the main mission hub at Cape Canaveral to computers located at the new Goddard Space Flight Center outside Washington, D.C. The four direct lines provided transfer rates just over one hundred kilobytes per second; a 4G cell phone today can relay data two hundred times faster. But at the time, the new link was blazing some thirty times faster than NASA's other options.[34]

If you grab a globe, it's easy to see the orbit loop chosen in NASA's early days; it was a circle that maximized its intersection with North America and, on the opposite side of the orb, friendly Australia, while minimizing any time over the Soviet Union. A reader may wonder why orbital space missions seem to fly in crazy curved paths, up and down over Earth. This sine-wave type of curve is actually just a flat representation of a path that completely circles the globe, a ring tilted away from the equator—we flatten that three-dimensional path so that we can see it on a two-dimensional map (see Figure 4.1).

FIGURE 4.1 The "big board" for early Mission Control. Without electronic means to display capsule locations, the capsule graphic was moved manually along wires. (NASA photograph.)

Cassetti was arguably the first American to experience an orbit of the globe. One of the early test missions collected a movie. NASA called Cassetti to the standard conference room where a technician set up the film projector for him again. "You're going to be the first person to see a view of Earth in color," the technician said. "I don't think there's ever been a color film of Earth." The empty orbiting capsule had held a movie camera pointing downward, and the film canister survived the heat of re-entry.

Young Marlowe sat down excitedly and prepared to take notes. "I thought, 'Oh boy! I get to watch the whole world!'" But the nearly ninety-minute silent film tested his patience. "The way you launch, you go over the whole Atlantic Ocean and then you kiss the coast of Africa and then you get the whole Sahara Desert. Half of the time, the first forty minutes, you're in daylight and you're seeing nothing. Just ocean or desert. And then you say, oh well I get to see Australia, but that's wrong because Australia is in darkness. And then you see Earth lighting up and you get the Pacific. And a quick run across the western United States and then you splash down in the Gulf of Mexico. . . . [I]t was really rather depressing."

The technician returned to the conference room, flicked on the lights, and asked what it was like. Was it amazing? "Water and dark!" Marlowe told him.[35]

This story presages something that would come to blunt America's space ambitions. As wonderful as the missions could be to the public, outer space could never quite live up to the human imagination and the science fiction stories that had primed Americans' interests. Nor could space compete with our buzzing, chirping, multi-hued home world.[36]

At the end of 1961, the communication challenges ahead were still daunting. Engineer Chris Kraft had been named to a new post, responsible for directing and choreographing the missions minute-to-minute after launch. It was largely Kraft dreaming up the architecture of what became Mission Control, that war room of consoles, headsets, read-outs, and cigarettes now familiar to all NASA fans. By Christmas of 1961, the agency had outlined phases two and three of the evolving space program. After getting single humans to orbit Earth (the major goal of the Mercury missions), phase two, Gemini, would send up two men at a time to test space suits against the hazards of space and to practice the dicey *rendezvous* of two separate spacecraft. Finally, phase three, Apollo, would somehow get humans to the Moon and back.

As of Christmas, 1961, the task ahead gave Kraft an involuntary shudder. "How the hell do we control a mission when the crew is a quarter million miles away?" he thought. "It's hard enough when the capsule is only 120 miles overhead."[37]

5

THE MOON

We can pause now in a way the engineers could not. We can relish a half-century's hindsight and appreciate the audacious goal itself. Humanity knew surprisingly little about the Moon in the early 1960s, but even today many of us could use a brief refresher on our closest celestial partner.

By borrowing and reflecting sunlight, the Moon has comforted Earth's evenings for billions of years. Our celestial dance partner's reliable cycle of crescents and disks provided a template for mapping humanity's sense of passing time. Its monthly procession from full to new simply broadcasts its location on its looping orbit. When farther from the sun than Earth (i.e., when it is "behind us" in the solar system's amphitheater), the Moon shows a brighter face. We see it much like we would turn in a darkened movie theater to see someone in a farther row, their face lit by the screen. When closer to the sun, the Moon's orb grows dark, as with someone's silhouetted head closer to a movie screen. And the half-disk confronts us when the Moon sits in the same row, as it were, with Earth.

Unlike a movie patron, the Moon always faces us. Like an obsessed admirer, our companion "phase-locked" to Earth's gravitational pull long ago. Gravity between any two objects pulls more intensely at closer distances. We tug on the Moon's closer side with a tad extra force—about two percent more—than we do its more remote side. Over time, this difference locked one side of the Moon to us as if we'd stuck it with an invisible harpoon. And this is a common affair. Many other moons in the solar system show the same gravitational devotion to their central planets.

The Moon, in turn, has a similar gravitational effect on Earth. More massive and stubborn, Earth tends to shrug off the uneven pull. But physics does not relent, and it slows the spinning Earth. We are, year to year and moment to moment, "phase-locking" to the Moon. Given enough time, one side of Earth, chosen by a slow physics lottery, would never see moonlight again, while the other would have the Moon ever present at one spot in the sky.[i] We grind toward that vision at a glacier's pace—tomorrow will be a tenth of a *microsecond* longer than yesterday.

---

[i] We don't have enough celestial time on the clock to actually have this come to pass. The inner solar system will be consumed by the sun's swelling red giant phase well before we phase-lock to the Moon.

Some scientists believe the Moon's pull played a significant role in nurturing our planet's complicated spray of life. Earth's spinning axis is more stable than that of the other inner planets, providing more regular seasons and climates over a much longer stretch of time. This arguably results from the Moon's perfect dance partnering. And compared to the solid parts of Earth, our malleable oceans more blatantly broadcast gravity's effect. They swell outward toward the Moon, with bulges rising and falling as ocean tides. In that way, by massaging shorelines and leaving tide pools with daily refills, the Moon may well have helped coax life from the seas long ago.[1]

In 1961, it compelled a newer species to consider a similar type of madness, enticing them from their natural habitat. But in NASA's early years, engineers saw little time or reason to contemplate that sort of ancient history. They needed to measure, compute, and troubleshoot a leap to the Moon. How can we digest the intervening gap? In a humbling, sun-centered perspective, the distance separating us from our nearest neighbor becomes absurdly small. If we could shrink our entire solar system to be an Earth-sized eight thousand miles across, Earth and the Moon would shrink by proportion to the size of dust motes. They would float together in the sun's glare just one centimeter apart, the width of a pinky nail.

But the separation looked substantial enough to the engineers. Any round-trip mission to the Moon would commit to some 480,000 miles, or about twenty times Ferdinand Magellan's distance in circling the globe. We can better grasp the distance as shown in Figure 5.1. In the figure, Earth and the Moon appear at appropriate

FIGURE 5.1 Earth and the Moon shown at proper relative scale but not with the correct separation. Note that this graphic breaks an important rule: The familiar side of the Moon facing the reader here should always be facing Earth.

scale: The Moon is roughly a baseball to our basketball. To separate the two appropriately, you would move this Moon graphic two-and-a-half feet away from the
Earth graphic—about five times the width of this book. In 1961, that gulf alternately
mocked and beckoned the engineers.

Assuming they could get a spacecraft to cover the distance, what did the engineers
know about the Moon in 1961? Even the best telescopes—all based on the ground and
forced to peer through Earth's thick atmosphere—could not divine lunar features
smaller than one thousand feet, about the width of a football stadium. Our best
images owned a panic-inducing fuzziness for those hoping to plan a safe landing.[2]

Faget's design partner, Caldwell Johnson, recalled these early days of the challenge. "I must admit, I didn't have a lot of confidence in the whole thing," he said.
"How in God's world are you going to ever hit the Moon in the right place that far
away? ... The scientific community was no help at all."

Indeed, scientists held little consensus on the Moon's surface itself. They had more
or less agreed that asteroid impacts, not volcanoes, must have caused most of the
craters. Yet, some observers had noted "ruby moonglow," bright winks of hot gas as
mysterious as they were temporary. Did they emerge after Moonquakes or perhaps
seep from hidden volcanic activity?[3] Some scientists believed the surface would be
too heavily cratered, rocky, and uneven to provide a legitimate landing place, and
still others sounded a different warning. "Nobel laureates would say, 'Hey, the thing
is nothing but a bunch of pools of dust,'" Johnson recalled. In fact, scientists debated
the lunar surface into the mid-1960s. They sifted through radar measurements,
computed temperature swings, and eventually used photographs from orbiting
probes, arguing over interpretations all the while. Some voices warned the engineers
that landing a spacecraft would risk having it slip into a dusty quicksand, never to
be seen or heard again. Others worried further that, even if an astronaut survived
the landing, he might stir up an ancient space virus in the lunar dust and, none the
wiser, ferry it back to a defenseless Earth. The engineers shook their heads and leveled
a more pragmatic eye on the Moon, deciding its harsh terrain probably felt a lot like
remote stretches of Arizona, but with conditions so brutal that no microbes or even
viruses could survive.[4]

Still, they planned any attempted landing with great caution. Early ideas for setting a craft to the lunar surface aimed for a slow approach, so the mission could
leap away if curb feelers found it too rocky or too soft. And NASA decided the first
landings would risk no lives. They opted to first make detailed maps using unmanned
probes (the Ranger program) and then test the surface with unmanned landers (the
Surveyor program).

Having not yet mastered laps around Earth, NASA began plotting orbits of the
Moon. Engineers balanced a number of factors. For a craft zooming away from

Earth, calculations would show what engineers called a "dispersion" of possible outcomes when it tried to hit its brakes and assume a serene, repetitive path around the Moon. The closer the orbit was to the Moon, the more those possibilities could lean toward violent impact craters. At the same time, an orbit too far from the lunar surface, while safer, required more fuel and power for a lunar lander to descend and return. NASA selected Apollo's optimum lunar orbit at a height of about sixty miles. A final piece of engineering logic and precision determined the altitude: An orbit of this size took two hours per revolution.[ii] Engineers would be able to look up at the humming, government-issue clocks in Mission Control and always know where the spacecraft was, without consulting some sort of timetable. (A similar logic placed many manned Earth-orbit missions in a path that took approximately ninety minutes.)

Even with a spacecraft sixty miles up, questions remained. Would the Moon's gravitational tug prove so varied and fickle that it ruled out a safe, circular orbit? And did the far side of the Moon have unexpected features, like enormous mountain ranges that could snuff a low-flying spacecraft? NASA assumed and hoped no geological feature could jut that far into space, but the engineers used to darkly joke about the sixty-one-mile-high mountain awaiting Apollo's first visit.[5] In 1961, the far side of the Moon still lurked as uncharted territory, like those blank patches of old maps filled with ornate, speculative sea monsters. The Soviet Union's Luna 3 probe had, in 1959, taken some very blurry photos of the Moon's unknown side. Despite its limited success, the probe still stands as a technical marvel. The earliest spy satellites (like America's Corona) needed to drop film packets back to Earth. But the Soviets couldn't return film canisters from the Moon at that point. No, Luna 3, never to return, developed its own film with an internal, automated machine and then used a primitive scanner to turn a resulting photograph into something that could be sent to Earth, dot by dot, via radio transmission. The final resolution was understandably terrible.[6]

Luna 3's images provided a Rorschach pattern: did a viewer see a welcoming world revealing its secrets, or did one see new complications and threats of failure? Far from easing any fears for the NASA engineers, Luna 3 and the Soviet space program served only as a kind of jittery fuel. The engineers stayed at their desks night after night plotting ways to shave bits of mystery from the Moon until they could hold it as a tractable problem.

[ii] Physics dictates that the closer a satellite gets to a star, planet, or moon, the faster it must move. Each and every orbital distance has its own required speed (independent of the size of the orbiting object). Closer orbits take less time, and higher orbits consume more.

6

1962—PUNCH CARDS AND A KEY
TO THE TRUNK

In a 1962 press conference, President Kennedy said the nation and the modern world were moving headlong into a post-political era. "Most of us are conditioned for many years to have a political viewpoint—Republican or Democratic, liberal, conservative, or moderate," he said. "The fact of the matter is that most of the problems . . . that we now face are technical problems. . . . They are very sophisticated judgments." In sketching a technocratic future, Kennedy said that modern solutions were "now beyond the comprehension of most men."

The idea that one could *engineer* solutions to thorny political and social problems had many people nodding their heads. The City of Oakland, California, invited a delegation of NASA and military officials to brainstorm city planning. They organized a conference for the following year, eventually sponsored in part by NASA, called "Space, Science, and Urban Life." Surely the problems of an urban landscape paled in complexity to the intricate workings of a massive rocket. They could be welded, computed, wired, and solved for good.[1]

More than half of Kennedy's 1962 budget went to military, surveillance, and the burgeoning space program. The CIA's first spy satellite, the Corona, had by then established that the Soviet military threat was more bark than bite. There simply weren't many missiles or missile bases. But Kennedy sought to make America the obvious model of choice for all the world's people. With his advisors, he also began outlining massive tax relief to super-charge economic growth.

Techno-milestones finally rose to feed this kind of giddy optimism. Early 1962 witnessed an uncommon victory for young NASA and a public outpouring not seen since the end of World War II. Riding skyward atop an Atlas rocket, an astronaut became the first American to actually orbit Earth. In a little under five hours, he would ascend, take three laps in a tiny capsule, and come back, barely, to Earth. "Nothing about John Glenn's flight was easy," flight controller Chris Kraft later wrote.[2]

Unlike the Soviet capsules of the time, NASA's first spacecraft had control thrusters: pint-sized versions of rockets that adjusted the capsule's orientation. And to highlight the fundamental principle of rocketry here, the smaller thruster rockets needed no flame whatsoever—just little puffs of hydrogen peroxide. Like the child on skates tossing a stone, the puffs gave the capsule a kick in the opposite direction. Early in Glenn's flight, one of the thrusters began misbehaving, moving the nose of his module left, like a shopping cart with a bum front wheel. But that was only the appetizer crisis. A warning light informed NASA that the capsule's all-important heat shield might be loose. An astronaut would never survive the fiery re-entry without the heat shield, sitting just inches behind his back. And as NASA opted to contrast their Soviet competitors by opening themselves to the press, Americans heard by radio and television that something might be very wrong with the capsule. The astronaut's return to Earth would be a tense and possibly deadly ride.

After some drama, with audiences waiting to see if a live astronaut would emerge from the recovered capsule, the nation enjoyed the sight of an upright astronaut, exhaled collectively, and then yelped with pride. In the end, the technical problem was simply a faulty sensor. Marlowe Cassetti says it was not much of a surprise. "It just really underscores in those days we were battling a lot of problems with the technology. . . . Nowadays an indicator comes on in the dashboard of your car or your airplane or your boat, you're pretty sure that there's a problem there, but in those early days, sometimes the indicators failed more often than the [systems] did." Moreover, the Mercury capsules were electronically messy. Engineers recall a crazy nest of wires running throughout its innards and weaving around the astronaut.[3]

In fact, the engineers had prepared the capsule, and the astronauts, for all sorts of sensor failures. Cassetti describes a set of physical *scratches* etched onto the capsule's window. The two most important tasks confronting any returning capsule were getting its wide end pointed earthward at the correct angle and initiating descent at just the right moment. Start seconds too early or too late, and the capsule's arc would splash down in unknown waters, far from the waiting recovery ships. Start minutes too early or too late and the capsule might hit with a deadly jolt on less-forgiving land. These last-minute adjustments were usually handled automatically. So, why the scratch marks on the window? If the automatic system failed, Cassetti says that the astronaut would "maneuver the vehicle to where those marks are on the horizon, then fire the rockets [to start descending]. That's pretty simple. It sounds pretty crude . . . but it's one that if all else fails and you lose contact with the ground and everything else, an astronaut, on his own, can line up that mark with the horizon, and he's got a stopwatch . . . so he knows it's just time and [orientation] and pull the switch."[4] Such a technique, like bygone mariners using a sextant, would eventually come in handy for one desperate Moon mission.

America's handful of astronauts had already achieved celebrity status by this time, but the euphoric response to our first American in orbit caught most people by surprise. We no longer seemed doomed to lag behind the mysterious and powerful

Soviet Union. And here was an idyllic, advertising-ready, crew-cutted American hero who boldly rode a dangerous rocket and fell in a meteor's trail to Earth, all to absolve the nation of its earlier failures. And the act was absurdly brave. The medical community had been genuinely worried—Gagarin may have orbited Earth and survived, but the Russians were keeping any medical information to themselves. Maybe the cosmonaut had returned half dead and could no longer speak or feed himself. Doctors sent a significant kit into space with Glenn: medicines to treat pain, shock, or motion sickness; and even shark repellant for his eventual time floating and waiting for rescue in the ocean.

His subsequent parade generated nearly 3,500 tons of confetti (still a post–World War II record). At a White House reception, the president's elderly father, Joe Kennedy, started crying when he met the astronaut hero, and he could not stop, even as his son tried to comfort him. "Now, now, Dad, it's all right, it's okay." It was anything but okay—a surprising new age was quickly flying over the old.[5]

Not every household had the same emotions for these early milestones. The Faget family had every reason to be jubilant—Max had devised a capsule that was taking men into space, where, now weightless, they could maneuver and measure and even marvel at Earth's serene curve through a small window. The capsule protected these men and guided them safely back without a scratch, a burn, or broken bone. "He'd bring home little models now and then that we liked," daughter Carol says of her father. "I still remember the model of the Mercury. You could hold it your hand." (See Figure 6.1.)

A quiet family scene unfolded at their two-story brick home in Newport News, Virginia, circa Glenn's triumph and not long before their move to Texas. Father Max was at work, and Carol was enjoying a carefree day upstairs. "At that age, I was probably reading a book . . . or playing with my Ginny dolls," she says now. "I came down the stairs and saw my mom on the steps crying. . . . Looking back, she probably did not realize I was upstairs instead of playing outside." Mrs. Faget had never been one to show much emotion in front of her kids.

A frightened Carol asked her mother why she was crying. "She was upset because my father wasn't going to be in this parade . . . that he was overlooked." Carol doesn't recall such a slight upsetting her father.[6]

Houston, Texas, embraced the arriving NASA engineers, and especially the astronauts, in their special Houston way. During a hot summer welcoming parade, astronauts waved to thousands of quiet onlookers, a crowd described by Tom Wolfe: "They stood there four and five deep at the curbs, sweating and staring. They sweated a river and they stared ropes. . . . They didn't even smile." Houston, to this day, offers sincere

FIGURE 6.1   Max Faget holds an early capsule model. (Public domain photograph originally printed in the pages of *The Virginian-Pilot*.)

hospitality and kindness to visitors, but in large groups, Houstonians are not easily roused to excitement. It may stem from a pre-air-conditioning culture, where people learned to move gently through monstrous summer heat. There's also the fact that, in local tradition, males abandon facial expressions by age fourteen or so.

Meanwhile, the engineers had no time for parades; they worked in a sort of roiling chaos. They hired waves of reinforcements, designed the facilities they would need at their new center, ran the first phase of the space program at full-speed, all while planning the next two phases. Engineer Aleck Bond's recollections speak for most. "Those were days when we worked ten, twelve, fourteen hours a day and sometimes seven days a week in order to be able to get the job done, and it was rather hectic at times. We neglected our families." And with their new campus just starting construction, they were spread all over, with rented space in shopping centers, bank buildings, a Canada Dry bottling plant, and, just off the freeway that headed south toward their empty pasture, a few addresses in the Houston Petroleum Center.[7]

The year 1962 saw great progress in NASA deciding *how* to approach the Moon; by year's end the dark horse plan had won. Not only would they use multiple craft, with necessary rendezvous of these ships in space, but NASA opted for having that take place *around the Moon*. The chief proponent[i] for this approach summed up its central appeal, having a separable, light lander, as follows: "I would rather bring down 7,000 pounds to the lunar surface than 150,000 pounds."[8] A number of central figures found the logic inescapable. Max Faget was one of the first engineers to change his mind, and he helped convince others. This approach provided the new freedom of designing a space craft that only needed to land and then lift itself from the Moon. It wouldn't need a heat shield. It wouldn't need to be sturdy enough to survive Earth's gravity and could therefore be *absurdly* light weight. It wouldn't need to carry the fuel required for a trip to Earth. All Earthly-needed things would stay up in orbit around the Moon, waiting for this secondary ship, a sort of metallic lunar insect, to gently alight below, look around, and return.

Wernher von Braun and his Huntsville team had originally supported a one-ship approach. They liked the simplicity and cleanliness of it, avoiding the need for *two* pressurized, habitable cabins, *two* systems of electronics, *two* sets of thrusters, and so on. They reasoned that having just one ship cut potential failures and problems at least in half. But they'd given that up once they embraced the reality of Kennedy's deadline. There just wouldn't be time to perfect a rocket powerful enough to lift one big do-everything ship away from Earth. Next they had embraced the plan with multiple pieces coming together in orbit around Earth, before embarking for the Moon, the big advantage here being that, if things went wrong, they could abort the mission close to home and safely return. By contrast, if ships had trouble around the Moon, perhaps even the far side of the Moon, with desperate astronauts blocked from contacting Earth, engineers would cling to few (if any) options for saving the mission. But von Braun's future ambition had also tilted him to favor assembling missions in Earth orbit. If NASA embraced such a practice early on, they could use the same method to prepare missions to Mars and beyond.

The various factions gathered for a crucial meeting in June of 1962, on von Braun's home turf in Huntsville. After various presentations, including some from his team advocating rendezvous operations near Earth, he quietly surprised everyone in attendance. The basic logic on the table had convinced him, and he announced support for the risky-sounding plan of a rendezvous operation near the Moon. Besides, this approach was the most forgiving for his Saturn rocket program. The lightest possible Apollo mission was one planning for the multiple ships coming apart and then docking again close to the Moon.

This simple, somewhat technical decision—rendezvous at the Moon rather than Earth—was a "seismic shift," according to Marlowe Cassetti. He'd been one of two

[i] John Houbolt was the most persistent advocate of "Lunar Orbit Rendezvous" (LOR).

Houston representatives at an all-day Huntsville planning meeting in the spring of 1962. He says the mood of the Marshall engineers grew somber as they saw the inevitable logic emerging. They recognized that a smaller overall rocket and a new lunar vehicle shifted more ultimate power to the Houston center and permanently away from Huntsville. Still, some of von Braun's staff held out hope, and they expressed some shock when their boss curtly changed his official position in June. Von Braun's incredible powers of persuasion knew when to yield. He was rarely stubborn when engineering logic sat before him.

With the need for orbital rendezvous now more or less certain, engineers had to push the idea from theory to practice, to learn its pitfalls and possible snags before attempting it in space, where lives and multi-million-dollar space ships would be on a collision course. And in Houston, where engineers shouldered all things spacecraft, they didn't have any facilities yet.

Engineer Tom Moser recalls an early improvised test facility. "We did it on an ice rink in the south part of Houston." On this borrowed skating rink, they slid two mock-ups of space ships around, trying to dock them together. Just imagine a grimly serious game of bumper cars, with a bunch of tired engineers scribbling notes on clipboards. In space, the engineers would face what they called "six degrees of freedom," meaning a ship could move in three directions (up-down, left-right, and forward-backward) but also *rotate* in three different ways.

If you, reader, hold your arm out in front of you, it's easy to demonstrate the three rotations, and their names, using your amazing shoulder joint. Keeping your arm stiff and pointing forward, first sweep your arm to point left or right—this is one rotation (called "yaw"). When you have your arm pointing forward again, rotate your entire arm to give a thumb up or a thumb down. This is a second rotation (called "roll"). Finally, as you alternately move your arm to either raise your hand or pat your thigh, this is a third rotation (called "pitch"). Floating in space, you then have six distinct ways to change the motion of a ship: three movement directions plus three types of rotation. When a craft engages several of these at the same time, mathematics gives the human mind one of its only grips on a literally dizzying situation.

Engineer Moser says it was still useful on ice with two dimensions of motion (everything but up and down), and one type of rotation (the left-right turning of "yaw"). But their test "space ships" didn't look the part. They were more like the offspring of wooden shipping pallets and primitive go-cart frames, each with an early piece of docking hardware tacked to one end.[9]

My parents made a move to Los Angeles in 1962, where my father started work on a vexing Moon problem: computing an exact travel path. Physics is very good at trajectories in general, and the discipline had two solid centuries of practice with artillery, but this Moon problem was something new. NASA needed to launch a craft from a spinning orb and have it hit another, smaller spinning orb at a distance of 240,000 miles—not *hit*, exactly, but rather slide carefully into a perfect orbit around that smaller body. Some engineers compared the task to aiming a rifle at a moving basketball many miles away, but to achieve orbit, the bullet would have to skirt the edge of the basketball by a fraction of an inch.[10] The engineers also had to reckon with an ever-changing tug on the spaceship. Earth's pull would decrease, mile by mile, once the ship embarked, and at some point a new pull would commence toward the Moon.

My father's new job involved working for Howard Hughes's aircraft company, as they contracted with NASA's Jet Propulsion Laboratory in southern California. This branch of NASA led, and still leads, the charge in *unmanned* probes. The overall plan called for robotic probes to orbit, photograph, and prod the Moon. Apollo could then find a safe spot to land with astronauts.

Shooting probes to the Moon had proven to be risky by 1962. Most of the Soviet Luna probes had missed their marks widely, if they even escaped Earth's gravity, and some had even crashed into the Moon unintentionally. America hadn't fared much better. The Ranger 3 probe had suffered a guidance system failure, and like a bad field goal attempt, it had sailed wide of its mark. And months later, Ranger 4's primitive computer had seized up on the way to the Moon, and with engineers wincing, the probe slammed into its target as braindead as a rock, learning nothing.

My father worked on the new Surveyor program, aiming to set probes onto the lunar surface. Hughes Aircraft, as the prime contractor, had a small team working on trajectories to the Moon. "I can't believe I used to do things like this," he says, looking at one of his papers from that era. The paper, with Robert Brown as the first of several authors, is called "The Generation of Lunar Trajectory Differential Coefficients using Patched Conic Technique," and its ten dense pages weave together mathematical equations and numeric tables. A "patched conic" is just a fancy way of suggesting that, mathematically, as a ship moves from Earth to the Moon, at some point we quit referencing its position using Earth as the mathematical origin. At some point, the ship is truly now in the Moon's *system*, its realm of gravitational dominance. But there is no simple way to change your equations from those using Earth as a zero point to another set using the Moon as a zero point. The entire ten pages of equations addressed this mathematical hand-off for the fledgling Surveyor program.

As retired engineer and trajectory mastermind Hal Beck explains, there was no clean and exact solution for finding a path to the Moon. Engineers didn't arrive at one equation, with one pleasing mathematical curve, for each trip to the Moon. Instead, they took a stepping-stone approach, where each stone was one little chunk of time on the way to the Moon. "It's hard to imagine, but . . . a lot of the work was done with a 32K [32 kilobytes] machine memory, and a lot of it was done with 64K, and that's almost no memory compared with even a small hand calculator [today]," Beck said. "When you would do long computations, you would take a time step and go through the computation and compute all the parameters [location, speed, heading, *etc.*] . . . and just print them." Then they would move their hypothetical spacecraft one more time-step forward, on its mathematical way to the Moon, and start all those computations again. "As a result, when you computed a lunar trajectory, you'd have a stack of paper like a foot thick."[11]

These early electronic computations formed a challenging bottleneck. "Machine time was so precious and lunar trajectory iterations were so consuming," Beck said. With the new Space Center still under construction, NASA engineers worked in their disparate locations and had to borrow computer time at the University of Houston. One competitor for such a precious resource was Langley transplant Marlowe Cassetti. While people like Beck needed the computers to help them find paths to the Moon, Cassetti was trying to compute exactly when to coax a Mercury capsule out of Earth orbit so that it would hit the Pacific Ocean at just the right time and just the right place. He recalls getting a taxi ride to the University of Houston's computing center during one of the frequent thunderstorms. Rainwater came spouting out from the sewers so fast "they were lifting the iron manhole covers off." Cassetti would gather his set of computer cards—the stack of carefully punched cards contained the program he needed to run, with the thumb drive still many decades away—and run through the rain. "You'd carry them in there, you'd submit them to a dispatcher, and then basically you don't get any results 'til the next day."

And he was often disappointed. "There was a high failure rate in those days," he said. "You'd punch one number wrong into a card, and you got a bunch of wasted paper, a big printout of garbage . . . one card out of order or one keypunch mistake . . . you wouldn't get any kind of diagnostic. . . . It would give you what was called a 'core dump,' just a bunch of octal numbers that didn't mean anything." Even when he'd punched all his numbers in correctly, he might be thwarted. "People don't realize, but in the early sixties, computers were tremendously unreliable. Very poor in terms of mean time between failures." In NASA's early years, failure was not only an option but also a common side dish to any project.

Beck, working on Apollo's exact lunar glide, was similarly frustrated, eventually calling this work—the stepwise, grinding computation for a trajectory to the Moon— the greatest challenge of his career. "It was so terribly sensitive," he said. "To tune that software and get it to function was a very frustrating job and took a lot of midnight hours," he said. And doubts left a pit in his stomach. "We were always faced with the possibility that it might not work." Even years later, when the first Apollo craft appeared to find its mark and slip into a perfect orbit around the Moon, he had to hold his breath waiting for the craft to appear again from the far side. Had their calculations worked? "It could come out at the wrong time, or it could be shooting off into space somewhere," he said.

Cassetti recalled a sort of unfortunate competition, given limited computer resources. The person running the university's computer center was a political appointee, according to Marlowe, and the fellow held little interest in the details of their work and deadlines. Under his watch, the computer center didn't seem to use much logic in determining which jobs got done first. "Occasionally I would tell our guys, we got to have this thing run by tomorrow," Cassetti said. "Wrap the cards in a five-dollar bill if that helps. It was so nonsensical."

But Beck had a secret computer technique, and his requests sometimes eased ahead of those from Cassetti and other anxious engineers: the trunk of his white Austin Healey convertible. "It was a sweet car, man," he says now. When dropping off his computer jobs with the night-shift university staff, "I'd give those guys a key to my trunk. I'd come back at six o'clock [in the morning] and have a big stack of results." The trunk hid an ice chest full of beer and a tacit understanding: the cans could disappear as long as the staff ran Beck's jobs (see Figure 6.2).[12]

The following year, NASA started buying more of its own computers. Engineer Ken Young recalled an IBM 1620 housed at the Houston Petroleum Center. At first, he felt lucky to have an office with just two people, instead of the normal three or four office mates. But their office sat next to the room reserved for the 1620, and even on the hottest summer days Houston could muster, the two engineers were uncomfortably cold. Mainframe computers need cool temperatures to keep themselves from overheating. The engineer and his office mate would sometimes try to sneak the thermostat up a few degrees, but they found a note from the head secretary charged with guarding this taxpayer investment: "The next person who touches this thermostat will be fired immediately."

Aside from borrowing and buying large early computers, NASA made a smart gamble on a new technology called an "integrated circuit," or a computer chip. The first such devices trickled onto the market in 1961 from Fairchild Semiconductor, and by 1963, NASA consumed about 60 percent of the company's chip output. The agency made a bold decision for the early 1960s: having Moon missions take their own computers along for the ride. These machines would need to be compact, lightweight, and reliable to an extent no computer had ever been. They would be violently jostled, zapped by space radiation, exposed to all sorts of possible temperatures, and operated by tired, stressed astronauts who were in no way computer experts. NASA tasked MIT's Instrumentation Laboratory with designing the Apollo onboard computers. And the Instrument Lab asked their young engineers if they could try to make such a thing work with these newfangled integrated circuits.[13]

FIGURE 6.2 Could you say "no" to this man's computer requests? Hal Beck shows off his Austin Healy in 1962. (Photograph courtesy Hal Beck.)

Just as engineers in Massachusetts worked on a newly petite scale of circuit architecture, another set of engineers, in Alabama, worked on rockets more massive than any before. The architecture of the eventual Moon rocket began to come together, at least on paper. A set of five new kerosene[ii] and oxygen engines, the largest ever of their kind, would power the first (bottom-most) stage of the Saturn V. And once that stage was exhausted of fuel and discarded, about forty miles in the air, the second and third stages would use something more novel: hydrogen-burning engines. In sum, the initial launch would be a brute-force affair, basically amplifying signatures of the German V-2 rocket, and then the next stages, moving into Earth orbit and beyond, would employ something futuristic, at least to the mind of young Marlowe Cassetti. "I said gosh, I can't believe they are building a hydrogen fueled rocket, almost like a science fiction type of thing."[14] But in 1962, both engine types were causing significant trouble.

[ii] With apologies to rocket engineers, I am simplifying by calling the fuel kerosene; it's a special blend called RP-1 that is fairly close to kerosene.

For the first stage, each new kerosene-burning engine (labeled an "F-1") would combine two tons of liquid oxygen and one ton of kerosene, *per second*, letting them burn at about 5,000° Fahrenheit, the temperature of a mild red star. To keep the combustion chamber from melting, the engineers used a clever trick from von Braun's V-2 rocket. Plumbing guided liquid oxygen, at –300° Fahrenheit, to fill a network of tiny tubes throughout the walls of the combustion chamber's bell, both cooling the chamber and also pre-warming the oxygen, before bringing it back up to meet its kerosene partner for their collaborative burn. But some voices warned that the ambitious F-1 engine might simply be too big, too thirsty, and too hot to ever work reliably. Fears grew in the summer of 1962. Rocketdyne, the company building the engines for Huntsville, ran a test in the remote Mojave Desert. They lit a prototype engine but watched it explode within a fraction of a second.

The most vexing problem for these mega-engines was called "combustion instability." Like a guttering candle flame, a rocket engine can develop an unwelcome pattern of fluttering burn rate, from bright and fast to dim and slow and back again, over and over. Once these pulses start thrumming in an engine, they usually don't stop, and the bell surrounding the combustion chamber can start vibrating like a chiming wine glass until it breaks. Even if the problem doesn't destroy the engine entirely, it can create what the rocket scientists named a "pogo" effect in honor of the bouncing children's stick. Like a car repeatedly lurching and braking, the engine instability creates a herky-jerky ride for the rocket, its equipment, and the eventual astronauts, endangering them all.

To further complicate the engineering detective work, two types of instability emerged: a lower-pitched variety that could arise within the structure of the rocket, and a higher-pitched version that came from the engine bell itself. Faget's boyhood friend and longtime collaborator Guy Thibodaux compared it to a musical instrument. "On an oboe, you finger," he said, to change the tone coming out of it. But "you don't change the frequency in rockets," at least not without significant cost and redesign.

Later that year, von Braun shared his worries in an internal NASA memo. He said the problem was assuming "new proportions" and that no single theory or idea could eliminate the instabilities. Engineers were left, he said, with an "empirical" approach—a trial-and-error game of shave this part, make that one thicker, poke an extra hole over here—which was expensive, dangerous, and, most of all, consuming time they did not have.[15]

Meanwhile, the second stage, built to take the Saturn rocket and its Apollo adornment through the upper atmosphere, was also providing major headaches. It's worth a bit of a deeper dive here, just to show the "hip bone connected to the leg bone" type of problems that plagued the Apollo program. Like the first, the second stage had five engines, but these relied on a technology that Marlowe Cassetti had viewed as Flash Gordon or Buck Rogers come to life. These engines would combine oxygen with hydrogen. As a very lightweight fuel, hydrogen gives rocketry the greatest bang for the buck. (Engineers talk of "specific impulse" for a fuel or an engine, which essentially measures how much kick it gives per gallon of fuel guzzled.) To be sure, using hydrogen created new headaches. To store it compactly within a rocket, engineers had to liquefy it, requiring a chilly −423° Fahrenheit. But the combination of its light weight and its incredible burning power was too great to ignore.

Hydrogen actually burns with a light we cannot see. Astronomy fans may pause and point out that the sun itself burns hydrogen, and is happily visible to us. But it's a completely different "burn"—the sun is using a nuclear reaction instead of a chemical one—and in any case, the sun is more like a hot poker than a flame. It sends out light based on its incredible temperature. The fact that we can't see hydrogen burning led to some unique quandaries for NASA and its contractors. In building and testing the new engines, engineers would occasionally encounter a leaking hydrogen line, and since hydrogen is so flammable, it could even start burning. You might hear it, but you would never *see* the flame. Bob Austin (he who recorded Sputnik's beeps and nearly destroyed his grandmother's kitchen making homemade rocket fuel) relays stories of a "broom test." If the engineers knew they had a hydrogen leak burning but didn't know where, they would move slowly along the hydrogen line or hydrogen tank, waving a broom in front of them. When the broom suddenly caught fire, they'd found their leak.

Having two super-cold fluids in stage II also created a novel design problem, especially since these fluids made up more than 90 percent of the stage's weight, when filled for launch. Super-cold liquids, if sitting in Earth's relatively warm atmosphere, boil and turn to gas, so we store such liquids in a fancy sort of thermos to slow the boil-off rate. They will keep turning to gas, but you can slow the bubbling to a crawl using good insulation. Stage II had 83,000 gallons of liquid oxygen, at −297° Fahrenheit, and 260,000 gallons of liquid hydrogen, at −423° Fahrenheit, and ideally they would each have their own separate insulated tank. But insulated tanks are heavy, and every extra pound on the second stage meant more lifting work for the first stage. The problems hit one another like dominos: if the second stage gained weight, and the first stage needed to do more work to lift its sibling, the final result meant the actual Apollo spaceship suffered mandatory weight loss. Its every system, including life support, would be thinner. The craft would simply hold fewer back-up options when things went wrong.

NASA decided (with their contractors at North American) to have the two super-cold tanks actually share a metallic wall between them. That decision saved a lot of weight, but the wall had to protect the colder hydrogen from the relatively warmer oxygen, or else the hydrogen would start boiling away before it could be burned. Engineers found a metal that could work, a special alloy of aluminum. Here, one of 1962's central problems unfurled like a roll of metal before the engineers. First, the size of these tanks was unprecedented for storing super-cold liquids. Next, the seams of the tanks needed the most precise and reliable welding jobs that America had ever achieved. And finally, the one metal that would work, the special type of aluminum, was notoriously difficult to weld. It would be one thing to do a tricky welding job for a few inches, or even a few feet, but stage II was nearly eighty feet high and one hundred feet around. The Huntsville team, and their contractors, had to invent new methods of welding and make sure these seams could then withstand a violent ride to space.[16]

Separate from welding the tanks, a glimmer of progress came from the avant-garde hydrogen engines themselves. Early tests had shown hydrogen burning with such fury as to melt most components around it, including copper pieces in the engine. But in 1962, engineers ran a prototype hydrogen engine for about four successful minutes without any part melting or exploding or fracturing. That might not seem very long, but the plan for stage II only needed it to fire for six minutes, roughly tripling the mission's altitude along the way.[17]

Meanwhile, a chance meeting in Huntsville solved an earthbound problem in 1962. A representative from a large equipment manufacturing company visited von Braun's outfit to discuss a crane they might need to stack rocket stages. But the rep overheard talk of a problem at the Cape: How would they move colossal Saturn rockets from an enclosed staging area to the launch pad, miles away? He thought his company had just the thing: vehicles the size of parking lots that stripped Kentucky soil away from seams of coal. These crawlers had tank treads, and, as he told the Huntsville engineers, the machines could self-level, meaning one could keep its roof completely horizontal, even though its treads were climbing a hill. What could be better for moving one of the world's tallest, heaviest objects?

One of von Braun's people went to see a crawler in Kentucky. After climbing on board and waiting for a while, the engineer finally said, "You can start!" But his coal-mining hosts said, "We've been moving for a couple of minutes now." Slow would be just fine for the job at hand, and NASA placed a custom order for a rocket-ready crawler.[18]

Lists of engineering head-scratchers weren't the only ones confronting von Braun and company. In August, as he approached Marshall's front gate, an expansive picket line greeted him. Electrical workers started the strike, but more than one thousand other unionized workers refused to cross their line. Tensions ran high as Marshall tried to build new test facilities at a whiplash pace. The strike started at a new monolithic structure, all iron and concrete, that aimed to someday provide a test stand for the first stage of the Saturn V rocket. The size of the new stand not only accommodated a rocket behemoth, but it also had to hold the monster in place during firing. Boasting as much concrete below ground as above, the new stand would eventually squat on the local wooded terrain like King Kong's own pizza oven. The priority and speed of the project led to an unhappy mix of union and nonunion labor with too much to do every day. Wernher von Braun complained to the press that the strike would cost America one million dollars per day and put us further behind the Soviet Union. The National Labor Relations Board stepped in after ten days, and work resumed.[19]

Von Braun witnessed another tension growing in Alabama. Civil rights leader Martin Luther King Jr. visited Huntsville in the spring of 1962, lending support to a series of sit-ins at segregated lunch counters in town. Several of these resulted in arrests. Just the year before, a white mob had badly beaten a biracial group of "freedom riders" in Birmingham, to the south of Huntsville. The freedom riders promoted racial equality by mixing races elbow-to-elbow on public buses (against local custom, to put it mildly). Von Braun spoke against racism as outdated, immoral nonsense, perhaps to further bury his Nazi past. But the Marshall Space Flight Center received two complaints from black employees that year. Appealing to President Kennedy's new Equal Employment Opportunity committee, the men complained of being denied promotion in one case and having inappropriate work assigned in another.

Wernher von Braun was developing a love-frustration relationship with his Alabama home. On the one hand, he liked to start his talks in D.C. and elsewhere by apologizing for his accent and pausing before explaining, "I'm from Alabama." Hearing a southern twang mix with a lifelong German cadence, one observer said the rocket scientist might as well adopt the middle name "Cornpone." But he had also started speaking up when he saw unfair views and counterproductive policies. For instance, he had a difficult time finding local workers with enough technical training. In 1961, he'd given a talk to the Alabama legislature, promoting investment in higher education. "Let's be honest," he told them. "It's not water, or real estate, or labor, or cheap taxes that brings industry to a state or city. It's brainpower." In the years to come von Braun would also speak plainly against segregation.[20]

Sometimes a significant loss for an organization can be as small as one person. With heavy hearts, supervisors in Huntsville watched Henry Pohl, the do-everything, do-anything engineer, prepare to leave. Pohl had new opportunities and new problems to solve in Houston, and the move would take him closer to his family. Maybe, he thought, he could even go work his father's cattle on weekends.

Transplanted Langley engineers were busily designing their new test facilities for the Houston site: they planned dedicated laboratories for testing a spacecraft's structure, its response to vibrations, its resistance to wild temperature changes, and how the plumbing of its engines handled extreme chemistry. They even planned a large "space environment simulation laboratory" where engineers could create the same conditions that ships and astronauts would encounter in outer space. Conscripting one of von Braun's Test Lab people made sense.

After Pohl's exemplary work on the small Saturn model, he was also now among the world's experts on tiny rocket engines. In particular, Henry had built his model rocket to fire inside a vacuum chamber that approximated the void of space. Now the engineers developing the Apollo spacecraft wanted someone to perfect sets of little thrusters that could maneuver in empty space, nudging the ship to its proper trajectory. Such thrusters would need to come on and shut off at a moment's notice, over and over again, with minutes, hours, or days between firings.

Henry had come to adore the environment at von Braun's center, with its many different labs and interesting minds. He particularly admired some of his forthright German supervisors there. Karl Heimburg, in particular, passed along some advice. "The day I left, he kept me in his office all morning," Pohl says. "It seemed he so badly wanted to see me succeed in Houston. He told me I was trained well. That no one there would know more than I did." Heimburg, born in 1910, had been a boy in one devastating war, an engineer in another, and then had crossed an ocean, half guest and half prisoner. He warned thirty-year-old Henry about the turbulence of a new organization under great pressure. He said Henry "would be like a speck in a pot of boiling water. At times the speck is on top of the water and then it is at the bottom."[21]

After the move, Pohl wasted no time in being his direct and plain-spoken self. "I had been in Houston about a week when the review of the first test facilities came up," he says. One of the main designers of the new facilities gave a presentation and asked for questions. "I said, 'Yes, I have a few, but first I want to thank you for a very good presentation. I know a lot of thought and hard work went into it. But there is a problem: it won't work.'"

His new boss chided him. "Henry, don't say that unless you can prove it."

Henry did. He described how the test engines would be placed too close to a wall and the exhaust would blow back from the wall and kill the engines. He set out the equations to illustrate the effect, and his boss, irritated, picked up some paper and went to work, double-checking what seemed like nonsense. His boss "sat there with a pencil and calculated it," Henry says. "Then he threw the pencil down and said, 'You are right.'"

Pohl had a budding reputation with his new colleagues in Houston, who "thought I knew a lot more than I did," he says. "I never did tell them that I had already made that mistake in Huntsville." But Pohl rolled up his sleeves and helped the team devise all sorts of facilities that would soon rise from the muddy pastureland.

He also needed new-found patience in Houston. In chaotic 1962, some meetings were more pragmatic and others more meandering and officious. "They would start with a list of the names of all the people that came in late and how many minutes they were late then get into the offsite telephone calls made and by whom and for how long. . . . After that they would go into all the mistakes they found in all the memos," Pohl writes now. "I thought that was a lot of trivia when we had so many pressing issues." Underlining his worries, NASA's Mercury program had logged about twenty hours of manned space flight at that point, compared to nearly two hundred hours by Soviet cosmonauts.

Pohl considered going back to Huntsville, but then began to see rapid improvements; he watched a NASA culture bloom in Houston where each subsystem or process would ultimately belong to one individual. Each buck would have a place to stop. And the engineering efforts broke into more and more sensible organizational branches. Henry joined the Propulsion and Power group, headed by Max Faget's good friend Guy Thibodaux. "Now this guy was much like the Germans in Huntsville," Pohl writes. "He didn't much care if all the i's were dotted or the t's crossed in the reports so long as they were accurate, and he had the best understanding of the physical sciences of any person I had ever dealt with."[22]

Sans Pohl, the Huntsville team kept pushing. To help resolve instability in the big engines, von Braun assigned one of his more experienced engineers to go to California to work directly with the contractors who were building and testing the engines there. "It aged me. I'm sure," engineer Jerry Thomson said of his trials with the F-1. By the end of 1962, the engineers were ready to accept the burn instability as a part of life. It seemed that it would arise from time to time no matter what. They could never predict when it would start, and no two engines were alike—the exact instabilities were essentially part of each engine's personality, and even then, one engine could have different moods from one week to the next. Sometimes the devils weren't even in the details—they seemed to move around between tests, finding ever-better places to hide. By the end of 1962, engineers started thinking about just making engines that could *survive* instability. If an engine, by design, could calm the staccato burning, if it could smooth out the problem over a few seconds, that might be good enough. NASA headquarters now worried that if this engine didn't shape up, there would be no Apollo program. Some voices started wondering if they should scrap that overblown V-2 type engine entirely and pivot quickly to a new design. And for Christmas, Wernher von Braun got two new lumps of coal: two more engines fell to burning pieces in the latest tests, as dollars and precious hours mixed in the flames.[23]

7

1964—OF DOUBTS AND BUGS

From my father's keepsakes, he pulls a small paper booklet published by NASA in 1964, bluntly titled "Why Land on the Moon?" and formatted for mailing, with a spot for a stamp. "You wouldn't believe it now," he says, "but people really had their doubts early on."

The booklet borrowed text from an article in the *Atlantic Monthly*, where two NASA scientists explained, step by logical step, why they supported Apollo and why the nation should as well. The authors listed positive outcomes to come for the many billions to be spent. In hindsight, some were more prescient (e.g., an economic boom in communications technology and a boost to science education) than others (e.g., providing the ultimate "Rosetta stone . . . even more important than Mars and Venus" for understanding the history of our solar system). But their main intent was as clear as it is now surprising: shielding Apollo from a barrage of critical arrows.[1]

On the heels of their troubled 1962, the negativity facing NASA and the Apollo program in particular had steadily grown. In the summer of 1963, the *New York Times* headlined a "Lunar Program in Crisis." Meanwhile, an internal NASA report told headquarters they had only a 10 percent chance of accomplishing President Kennedy's end-of-decade goal. Former president Eisenhower poked his head up from retirement to say he feared the burden of a Moon program on the nation's taxpayers.

Even the nation's scientists were taking shots. The editor of *Science* magazine penned a common view: The same scientific knowledge could probably be gained for 1 percent of the cost, if the nation would just go with an *unmanned* approach. In fact, many scientists accurately saw Apollo for what it was and what it would ultimately be: more a proof of technological principle than a mission of scientific discovery. And in terms of cost, it was clear that NASA didn't just need to climb a high engineering peak—they would need to build several new bridges just to get to base camp.[2]

The scientific community felt themselves on the outside looking into an agency that, early on, employed twenty-six engineers for every scientist. External groups of scientists worried that Apollo would drain NASA funds away from scientific missions. They listed their scientific priorities for the Moon, such as uncovering its geology, probing its interior via surface instruments, and measuring the ephemeral lunar atmosphere. While many of these eventually did come to pass, many scientists felt ignored, and the tensions were real enough. "Scientists and engineers simply do not think alike," engineer and flight director Chris Kraft later wrote. "One of the important lessons [of his time at NASA] was that any apocalyptic prediction by a scientist would almost certainly be wrong."[3]

Even the selection of landing sites became contentious in the early 1960s. One leading scientist, Harold Urey, said he could not tell who was ultimately making the decisions for where to land, and when he *did* learn of the sites, the communications were so chock full of engineering acronyms as to be unreadable. Engineers did consider the scientific requests. Lighting made for one incredibly important factor, and not just for photo opportunities. Landing at a site and time where light was plentiful but also falling at a good angle became a central requirement. If the sunlight hit from directly overhead, the lack of shadows would obscure features from the view of the astronauts trying to land (and from photographs that could help geological scientists understand the Moon), but if the shadows were too long, late in the month-long lunar "day," too many features would be in complete darkness. Unlike a shadow on Earth, a lunar shadow is truly dark, with no atmosphere to introduce an ambient glow. Engineers would take the scientific requests and put those through their trajectory paces. "All the landing sites are within a fairly narrow band of the equator," said engineer Hal Beck. "That's where your basic orbit was, and so you couldn't get up to higher latitudes." Given the tight squeeze on every pint of fuel, an Apollo mission didn't have much leeway for exploring the lunar map.[4]

Meanwhile, Khrushchev cooked up new PR challenges for NASA as well. The summer of 1963 saw cosmonaut Valentina Tereshkova become the first woman in space. "Bourgeois society always emphasizes that woman is the weaker sex. That is not so," Khrushchev crowed. "Our Russian woman showed the American astronauts a thing or two. Her mission was longer than that of all the Americans put together." Official Soviet statements would soon return to typical male-centered rhetoric, with Tereshkova herself saying, "[N]o work done by a woman . . . can enter into conflict with her ancient 'wonderful mission,' . . . the bliss of motherhood." But Americans had found a new way to shake their heads. We didn't just have a missile gap now. One Senator's wife suggested that America was "one hundred years behind in using the full abilities of women." Indeed, the engineering workforce in the Soviet Union enjoyed a much higher percentage of women than that in the United States.[5]

Interestingly, the U.S. space program had started training female astronauts and found they physically stood up to extreme conditions just as well as, or even better than, their male colleagues. More than a dozen women had excelled in early 1960s training programs. (As of this writing, we know that women are not just the equals to men in space, but they may be better suited for long-duration voyages in deep space.) When Vice President Johnson saw a summary of the early results, however, the tradition-bound southern gentleman scrawled "Let's stop this now!" across the top.[6] The first American women would have to wait.

Late in his abbreviated presidency, Kennedy had actually made public overtures to the Soviets, hinting at (and even presaging, we can say now) cooperation in space. After announcing a nuclear test ban treaty with the Soviets, and then installing that icon of all Cold War films to follow—the "hotline" phones linking the Kremlin and the White House—Kennedy had told an autumn 1963 United Nations assembly, "Let us do the big things together." Khrushchev's son later revealed that his father continued to warm to the young American president and was even receptive to the idea of joint space exploration. By November, however, Congress had cut over $600 million from Kennedy's budget request for NASA and even included a rider ruling out a joint lunar landing with the Soviets.[7]

Kennedy's tremendous physical discomfort had only grown since his election. By late 1963, he spent a great deal of time in a heated pool. But he had maintained his support for NASA and a presidential travel schedule, including tours of space centers. He took what would be his last trip to Cape Canaveral on November 16, meeting with various NASA leaders on a tour of the burgeoning Apollo facilities there. (See Figure 7.1.)

FIGURE 7.1 Wernher von Braun leads President John F. Kennedy on a tour of the Marshall Space Flight Center in September 1962. (NASA photograph.)

Just a week later, a startled nation had lost its president to assassination. November 22, 1963, had been set as the date American satellite technology would transmit the first television signal across the Pacific Ocean (from the United States to Japan): a pre-recorded greeting from President Kennedy. With his jarring, public death hours before the broadcast, officials quickly scuttled the event. A few days later, the new communications satellites (or "comsats" to the engineers) beamed footage of the presidential funeral around the world. Lyndon Johnson, newly sworn in as commander in chief, named NASA's Florida launch center after John F. Kennedy.[8]

Within a year, a less dramatic shift in the space race arrived as a quiet Soviet coup removed Nikita Khrushchev from power. While his chief rocket designer Korolev would continue struggling with a new, more powerful booster rocket, the incoming leadership began to emphasize scientific discovery over flashy missions that could embarrass their American rivals.[9] But Korolev was left with a specific Khrushchev order that he didn't relish. Mindful of America's Apollo plans, Khrushchev had ordered his chief designer to get a three-seater into space, sooner rather than later. Without time to properly develop a new spacecraft, the Soviet engineers simply removed the ejection seat and reserve parachute from their working one-seater (the Vostok) and announced a "new," crowded craft (the Voskhod) ready for three cosmonauts.[10]

America had lost its president and NASA had lost its political prophet, the man who had pointed at the Moon and made a promise to Earth. The agency's setting for 1964 looked especially dark. Books like *Moon-Doggle* and *The Rise and Fall of the Space Age* made loud pronouncements of Apollo's coming demise and a nation's science fiction folly. And additional worries plagued the sleep of NASA's leading rocket architect. A book had emerged in East Germany titled (in translation) *Secret of Huntsville: The True Career of Rocket Baron Wernher von Braun*. The book uncovered many unflattering facts of von Braun's Nazi-era past and his compromises with the Third Reich. Von Braun was concerned enough that he warned NASA leadership, but they told him to ignore it. Despite some legitimate content, the book also owned a breathless, tabloid tone and embellished many stories with hyperbole. Given America's well-founded mistrust of Soviet-bloc propaganda efforts, the local press ignored the book, and it remained hidden in the German language and behind the Iron Curtain.[11]

In public, Wernher von Braun led the agency's attempts to counter the negative space narrative. He had entered a full-court press of public relations. In this era, he gave and wrote around 150 major public speeches and articles *per year*. These included monthly columns in *Popular Science*, and he shrewdly confronted other topics, putting our space missions in the context of all things America: democracy versus communism, for one, and the dialogue between science and religion for another. Witnesses describe a public confrontation between a church deacon and von

Braun. The deacon complained of an Alabama drought and accused von Braun of "punching holes in the clouds with those rockets and drying up the rain."

The rocket scientist deftly swung to the Bible and reminded the audience of Jacob's ladder, with angels ascending and descending. He said that humanity was now on the ladder, taking the first steps. "If the good Lord does not want us to go up and down His creation, all He has to do is tip over the ladder." The crowd erupted with applause.[12]

In the face of national doubts, NASA leaned on the strident political support of Lyndon Johnson, who had originally huddled with von Braun and recommended the Moon missions to Kennedy. Shortly after becoming president, Johnson told an economic advisor, "To tell the truth, John F. Kennedy was a little too conservative to suit my taste." He put a team to work on a major anti-poverty initiative, and the drive to the "Great Society" was on. If Kennedy put the nation's do-everything ambition in high gear, Johnson floored the accelerator pedal at home and abroad. As he famously said in 1964, "Surrender anywhere threatens defeat everywhere." Selecting from a menu of policing the world against communism, rebuilding impoverished neighborhoods in America, or engineering an expensive trip to our nearest celestial neighbor, Johnson voted for all of the above.[13]

Of all the national voices, an inhuman one spoke loudest to Wernher von Braun, as the singing of the F-1 engine resisted all coaching. Given von Braun's outreach work, it would be easy to assume he stayed well above the technical fray of the Marshall Space Flight Center, but he did most of his public writing at night. Even though he had excellent managers working under him, he maintained his vital role for realizing the Saturn V. At the start of 1964, the hazardous warble in the immense new engines remained Huntsville's chief worry. Combustion instability threatened the whole enterprise, still striking without warning and too often to ignore.

As the engineers thought less about complete elimination of the problem and more about building an engine that could dampen the effect and survive it, they started to regularly *induce* the unwanted vibes in their tests. They sought to create the worst warble imaginable to make sure a firing F-1 engine could then tamp it out or at least remain intact. In a surreal move, they began setting off small bombs within fully burning engine bells. The blasts would start up an uneven burning pattern. Some had predictably catastrophic results, but when an engine survived and returned to normal, smooth burning within half a second or so, that marked success.

Two contract engineers working on these tests found themselves simmering in a new kind of heat. During a commercial flight, the pair argued about their bomb tests. A flight attendant overheard their energetic discussions and grew worried. When they deplaned, dark-suited FBI agents escorted the engineers to an interview room and proceeded to grill them for hours about their suspicious bomb discussion. Their story—we're setting off little bombs to help test rockets for the Moon mission—struck the agents as absurd, and eventually top brass from the engineers' employer had to intervene.[14]

At the peak of instability panic, 125 engineers and some 400 technicians focused on the problem. And 1964 did witness progress. One type of instability responded to an ad hoc solution that tweaked existing engine structures. Fuel entered the F-1's blast chamber through, in essence, the world's largest showerhead, with some 3,700 holes for the kerosene fuel and another 2,600 for the oxygen. Engineers began altering the angles of these holes through the plate, giving a little more variety to the directions by which fuel would enter the blast chamber, and they also introduced small metal ridges on the blast side of the showerhead. In this case, the engineers had guessed right: These "baffles" helped dampen the instability when it arose, perhaps by interrupting spiral flows of burning gas.[15]

Engineers also worked through a different kind of instability, and at least in this one case, they actually found the culprit. "There were ten or twelve of us in an office that was twenty by forty feet," Huntsville engineer Len Worlund said. "We were all sitting there, desk to desk in two rows. I could turn around and bump an elbow on the desk behind me. You knew everything that was going on." When they faced a tough problem, they would sometimes start developing three possible solutions in parallel, so as to not lose calendar time.

The engineers, predating computer screens to display their measurements, printed everything. In probing one of the instabilities, they tracked the pressure inside a fuel line over a long engine burn and printed out the graph. It wouldn't fit on their desks. "We'd unroll that oscillogram down the hall," Worlund said. "You'd see a little bitty trace oscillating." They carefully marked the time between the warble's peaks on the printout, and in time they uncovered an odd sort of communication between different structures of the engine. One engine part (a long metal pipe transporting liquid oxygen) and another (in the wall of the engine's bell-shaped blast chamber) happened to ring at the same low tone, just under the range of human hearing. When the thin pipe started singing, the engine bell couldn't help but join in. By year's end, the engineers thought they'd handled the problem by restructuring the fuel line just enough to change its tune. (This instability villain would lurk, however, awaiting the moment a Saturn finally vaulted into the air.)[16]

Worlund, then embedded in the instability issues, describes flights with von Braun from Huntsville to the Cape, sometimes for a presentation and other times for a test launch. "Von Braun always liked to take off, so he sat in the copilot seat." But once they were airborne, he moved to the rear of the plane to start working. Worlund recalled one intense trip, flying to update staff at the Cape on the instability problem.

"And he [von Braun] said, 'Let me see your charts.' He's sitting there and flipping through those charts, and he says, 'Guys we can't do that. It'll never fly.'" Von Braun wanted to find a more optimistic message for their technical presentation, and they reworked the game plan in the air.[17]

Instabilities weren't the only problems von Braun's team faced in 1964. An alarming number of cracks emerged in most of the test engines. The team tracked each one like a murder mystery: collecting and analyzing evidence until they could convict a perpetrator. A strange act of chemistry created one set of cracks, with components of exhaust combining and conspiring, like bad influences, on the metal walls of the combustion bell. In other cases, the super-cold liquid oxygen simply froze and cracked the engine materials. Engineers alternately turned to new materials or devised special pre-treatments to strengthen metal components against the extremes of fire and ice.

Meetings at the Marshall Space Flight Center regularly ran past midnight. Many of these were coordinating sessions, just to keep track of how solutions to one problem would affect other segments of the rocket. Some describe "a battle" to keep track of daily changes and follow them logically—knee-bone to thigh-bone and so on—to their predicted impacts in other systems. This hive of details buzzed within a carefully orchestrated set of well-worn file folders. Engineers describe conference rooms thickening with cigarette smoke as meetings paced into late nights. Von Braun was an ideal leader through such sessions, attracting respect and inspiring excellent work. Colleagues look back and admit they never hesitated to work around the clock for him.[18]

Many engineers in both Huntsville and also Houston recall an extra task that was rare in other organizations. "Every single day they required every engineer to write one page, a handwritten page, and send it in to your boss, of what you did that day," Henry Pohl said. "The section [leader] took that from all of us and condensed it into one page that was passed out the next day to the branch, the next day to the division, and the next day up, and every day von Braun got from every laboratory a one-page summary. It always stayed one page, but it moved up the line." This was not just an exercise, as Pohl learned; supervisors scrutinized these summaries. He once received an early morning phone call from an irate manager. Henry had written that he followed normal protocol on handling a fuel tank but then admitted he could see a method less prone to mishap. "I mean he just chewed me up one side and down the other side," Pohl said, "for agreeing to do something that I knew was not safe."[19]

But work for von Braun was not all smoke-filled conference rooms and reams of one-page work summaries. In 1964, he and a few of his long-time German colleagues became regulars at a new lounge, the Top Hat. The owner later recalled von Braun showing up in the gravel parking lot around 5:00 p.m., and then seven or eight "other doctors" followed. The group, she said, would "head straight for our stockroom in back, where all the beer was stored. Only American beer. They would drink the beer hot in there and talk and draw all over the beer cases—rocket designs and things." After about an hour, von Braun would invite her into the room to survey their empty bottles—neatly arranged for counting—and compute a bill. She said this weekly routine spanned most of the Apollo years. She regretted never saving the cardboard cases that mixed German rocket scribbles with American beer logos.[20]

Von Braun started to see reason for optimism in 1964. He promised NASA headquarters that he would boost the power of the Saturn V rocket, until it could lift an extra half-ton into space. With the mission featuring multiple modules now, every pound counted, and von Braun said his team could lift a thousand more.[21]

In the fall, von Braun got a special treat, seeing part of what his rocket would eventually heave toward the Moon. NASA headquarters invited him to climb into a first-generation "lunar module," a light, insect-like landing craft that would separate from the other Apollo modules. He had always enjoyed flying and wished he could actually take a mission into space. In the Mercury program, he joked that he'd been declared too fat to join the astronaut core. But in 1964, this was his first genuine tour of a ship designed specifically for foreign worlds. After carefully inspecting the innards of this Moon lander, he emerged like an excited child. "You've got to go up there," he called to a colleague. "It's great!"[22]

By now, support for the lunar lander plan was widespread. "We gave it a good hard look," Max Faget later said. "Not only did it solve the problem of being able to get there with one launch of the Saturn . . . but it solved problems that we didn't think about before. . . . [It] didn't have to have any heat protection, no aerodynamic considerations at all."[23] In 1964, NASA set out some very basic requirements for the lander. It would need to ferry two astronauts from lunar orbit to the surface, along with 250 pounds of extra equipment to be left on the Moon. The craft would land on the side of the Moon facing Earth, with an antenna pointed aloft to maintain radio contact. And it would need to then lift roughly one hundred pounds of lunar soil and rocks back to lunar orbit.

Initial ideas for this nimble little Moon craft varied widely. One early sketch showed just a skeletal frame holding one standing person. While that fanciful idea never really had a chance, it did prove prescient in one respect. Engineers struggled to figure out a way for seated astronauts to effectively see the Moon's surface using a limited number of tiny windows. (Windows were extremely heavy and also had to be perfectly sealed against the vacuum of space.) Engineer Thomas Kelly, one of the leading designers, described the flash of insight that both saved weight and improved viewing angles. "What if we get rid of the seats?" The final lunar lander was a standing-room-only affair, with tethered astronauts standing upright and using the small triangular windows. In fact, given standing astronauts, engineers decided to make the windows even smaller, saving more weight. And losing the bulky, cushioned seating created extra space in the tiny craft as well.[24]

Having a craft that would never have to deal with Earth's atmosphere or Earth's strong gravity gave designers a vertigo-inducing freedom. They debated whether all the key equipment (life support, storage tanks, communications, etc.) would be best placed inside, with the astronauts, or outside, on the hull of the craft. If it didn't have to worry about the friction of an atmosphere, this ship could be lumpy. Rough-draft models proposed a spherical crew chamber, sort of like a diving bell. The lander would have the most space for the crew using the least amount of metal if it was round. The little promotional pamphlet provided by my father featured cartoon sketches of this early prototype landing on the Moon. With its spherical head and two little cylinder-shaped docking hatches, it looks like a cartoon pig wearing a fez. Many changes were still to come. Eventually, engineers opted for a lumpy exterior and a more cylindrical crew chamber. "It didn't have to be pretty," recalled Faget, laughing. "It was nicknamed 'the bug,' which everybody objected to . . . but it looks like a bug."[25] (See Figure 7.2.)

FIGURE 7.2 A dreamy, early diorama of a lunar lander. (NASA image, via Grumman.)

To save every ounce of weight, engineers planned to shave components to be as thin as physics would allow. As long as it could hold air within it for a few days, like a sheer metallic balloon, that would suffice. All the pieces, even the plumbing of fuel lines, were shaved, etched, and thinned to save weight. As early as 1964, however, engineers found worrisome cracks in the spider-like aluminum legs.[26]

The initial requirements for the lander's legs were as mysterious as the nature of the Moon's surface itself. Engineers moved from a five-legged critter to a four-legged one, with large platter-like feet at the end of each spindly leg. If the lunar surface was dusty and soft, the large pads—each about the size of a banquet serving tray—would keep the lander from sinking. And if the surface was relatively hard, like a parking lot, the large feet would just add stability against tipping over. But if the Moon was *too* hard and we couldn't bet on a gentle landing, the legs would need shock absorbers. Normal shock absorbers involve some kind of fluid, but this adds weight, and fluids could easily leak into space, freeze in the cold, or boil in a blast of sunlight. Engineers arrived at a fluid-free system: a thin honeycomb structure that would softly collapse under the pressure of landing. It could only be used once; the lander could not take off and land a second time.[27] In fact, to simplify the return trip, the bug would molt: its legs and landing structure would stay on the Moon, even as its upper half rocketed back into lunar orbit.

That leap off the Moon presented one of the Apollo program's most worrisome segments. In talking with the engineers, you hear a mantra of "fail safe" systems; nearly every segment of the mission planned to have a back-up method in case some piece of equipment failed, and most segments had a back-up to the back-up. But there was an exception: the one, lone engine that needed to lift the astronauts back into space. "To me, that was the biggest problem point," my father says. "If it didn't fire, you were dead." Other engineers concur that this "ascent engine" was a rare "single-point failure" for the entire enterprise, with no back-up plan. If that engine didn't fire properly, NASA would maroon two astronauts with a few extra hours' worth of oxygen, and the world would watch them slowly asphyxiate next to the American flag.

When a single piece of gear is this important, engineers remove as much complexity as possible. As lander designer Tom Kelly wrote, "It was simpler than the common oil burners used for home heating." This one didn't have special fuel pumps or intricate plumbing or the ability to aim the direction of its exhaust, like the Saturn V main engines. "This innocuous-looking device," Kelly wrote, "proved to be one of the greatest threats" to realizing the Moon missions.[28]

The prototypes were also dangerous on Earth. One way to simplify an engine is to eliminate an *ignition* system. Like a stove burner, most engines require an initial spark to get going, but we've all seen stove burners that quit sparking—they just emit a hiss of useless gas. An alarming but useful alternative involves using "hypergolic" fuels, two or more chemicals that start burning as soon as they mix. These had been used before in missiles (like the Titan series), and other modules of the Apollo mission would use them, on a smaller scale, as thrusters that let a ship rotate itself or tweak its course. (Henry Pohl, master of small rocket engines via Huntsville, would be working on those for the main Apollo modules.)

But the big ascent engine for the lander dwarfed those thrusters, and even though it faced only the Moon's weak gravity, it still had to lift itself, two astronauts, and one hundred pounds of Moon rocks back into space. The two chemicals of choice here, giving a great bang per ounce of fluid, happened to be extremely toxic.[i] One, when leaking, formed a cloud-like white vapor, while the other was a more earthy, brownish red. Checking for leaks in early engine design was critical not only for future mission survival, but also for the lives of the Earth-bound staff putting the system together. One engineer recalled dipping a stick into one of the chemicals and letting it drip onto snow outside his laboratory, only to watch the snow burst into flames. Another story from these times had a distracted technician click a ballpoint pen against one of the paper-thin storage tanks, where he unwittingly created a tiny leak. The incredible pressure shot the pen, with a good piece of his finger, into a nearby fence post.[29]

Before a lander could launch back into space, it would have to navigate its way to the bumpy lunar surface. Robert Brown accepted a job with rapidly growing NASA in 1964. His work experience at Hughes Aircraft looked like a good fit for the Apollo missions. Not only did the unmanned Surveyor program require all the calculations of getting to the Moon and making adjustments along the way, but Surveyor also aimed to land softly on the Moon, where it could take photos and make some simple measurements. Did they use some sort of remote means, like a mission-control joystick, to land the Surveyor probes? No—it was all automated. The Surveyor flew solo, using no astronauts and no remote human input, to land on a foreign world of largely unknown terrain and texture. "It was very simple," my father says of Surveyor. "We *had* to make it simple." Surveyor lived on a fraction of Apollo's budget and a relatively small team of engineers. A radar unit on its underside aimed at the Moon's surface. The radar was actually *inside* the cone of its landing engine. When the craft got close enough to the Moon "the radar triggered the descent engine and blew the radar out." And destroyed it? Yes. "We didn't need it at that point."

As of 1964, the planning and a lot of the construction for Surveyor were complete, but it had not yet flown to the Moon. Still, weren't the lessons learned attractive to Brown's new employer, NASA? "No," he says with a tinge of sarcasm. "They knew better, on everything." Inside the agency, the engineers were (despite or maybe in

---

[i] Aerozine 50 and N_2O_4.

concert with their youth) developing an iron-clad technical confidence. If the nation had any expertise for space travel, they reasoned it had to be in-house.

Records show that the Soviet space program had already embraced automated systems for most of their space journeys. The cosmonauts did not have much to do. Even when they eventually docked two ships together in space, the humans primarily enjoyed the show as computers did the work. But for the American space effort, the astronauts maintained a more central role. Later in the Apollo program, worries about astronaut safety intensified. "We put a group together to see if we could fly unmanned," my father says, "and sure enough, we figured you could do it. But that was nixed also. The question comes up about why you have all these astronauts."

In late 1964, the lander program had plenty of stressors, with or without astronauts on board. In addition to cracking materials and a dangerous liftoff engine, its thruster prototypes—the smaller rockets for maneuvering—sometimes backfired and exploded. And unlike the unmanned Surveyor, the lander needed more complicated radar systems. It was one thing to set down a hunk of metal on the Moon, but it was another thing to set it down gently enough that a standing human being wouldn't break his legs. Plus, while a Surveyor stayed put (five of the seven probes landed and will sit on the Moon for eons to come), an Apollo lander needed to shoot part of itself back into space, find the rest of the Apollo spaceship, and carefully dock with it. Having radar for the Moon landing *and* for docking added weight, added complexity, and even generated unwanted heat. Moreover, engineers weren't sure the technology could provide good enough information to the astronauts. NASA actually considered removing the radar systems and trying to inform the Moon landing and the rendezvous of ships using large Earth-based communication systems. This didn't come to pass, but imagine it for a moment: two fragile ships orbiting the Moon and awaiting moment-to-moment orders from Earth. "A little to the left . . . no, your other left." And all these signals would be slightly delayed by the length of two astronaut heartbeats, the time for radio waves to speed a quarter-million miles.

When a rough-draft, full-scale lander model came together in October of 1964, Wernher von Braun had been one of the first and most enthused humans to crawl inside, but the model showed a new problem and prompted changes right away. A worker played the role of an astronaut trying to get in and out of the thing, but the exit hatch was too small. A suited astronaut with his life-supporting backpack could only squeeze through after superlative effort, risking suit damage.[30]

The story of the space suits illustrates the equal importance of every level of detail, from titanic rockets to tiny stitches meant to protect a man from the harsh indifference of space. Like a self-destructive engine, or a problem-filled lander, the suits themselves made for another distressing issue. A test of the leading Apollo suit candidate failed in spectacular fashion in late 1963. Staff, playing the role of astronauts, could not rise from prone positions—the awkward suits pinned them on their backs, according to one account, like a bunch of helpless turtles. In the spring of 1964, NASA took a rare step and declared "contractual failure" against the suit maker, Hamilton Standard.[31]

Early in the space program, some had considered a hard suit, like an insect's exoskeleton, as the most sensible model. The joints were always a difficult matter, but storage was just as damning for hard suits in a Moon mission. A hard suit was like an extra passenger in a mission with no extra seats, but a soft suit could be folded and compressed when not in use. (The hard suit idea persisted in the background, and in a more secret space program, the U.S. military preferred hard suits when planning an orbiting outpost. The proposed space garrison would have had a dark-room to develop spy photographs of the USSR, but the idea lost out to unmanned spy satellites.)[32]

Any spacesuit for Apollo faced a decathlon of space requirements. It was one thing to build a leak-proof suit that could keep a man alive in the vacuum of space. The extreme temperature differences of the light and dark sides for any object floating in our solar system posed another challenge, with one side of a suit being hotter than boiling water, and the other colder than any deep freeze on Earth. But even if a suit reliably overcame these hurdles, the suit then needed to allow people to *work*: to move, to lift, to grip, and to press keys. Astronauts would need to operate equipment, climb down and then up ladders, and possibly survive falls onto sharp Moon rocks.

An unlikely dark horse challenged the larger suit-making contenders. The company that brought "Playtex" undergarments to America, the International Latex Corporation, was now making their best pitch for NASA's Apollo spacesuit. The Latex Corp. had worked on specialty pressure suits since it first won a small contract with the air force in 1950. NASA originally had them subcontract to Hamilton Standard, the presumed favorite contractor for making the Apollo suits, but that collaboration started to unravel in 1964. While Hamilton was a button-down, formal defense contractor, their smaller partner was a different animal—less hierarchical and more averse to paperwork.

The Latex Corporation encouraged the input of the seamstresses, some of whom were pulled from their work on diapers. These women deeply understood the various fabrics and materials of the many layers in a spacesuit and knew what it would take to stitch a seam so flawlessly as to make it airtight. The final Apollo suits, fit for a walk on the Moon, gathered a total of twenty-one layers. In their Delaware offices, managers and seamstresses huddled in late-night sessions. Over coffee and cigarettes, they debated materials, fabrics, and the ordering of the layers, balancing safety versus flexibility on the top of a sewing needle.[33]

Half a continent away, NASA managers stitched together another complex structure, layer by layer. New hires and Langley transplants streamed into NASA's new Manned Spaceflight Center south of Houston.

My parents moved to the area in the spring of 1964, as my father joined the "Mission Planning and Analysis" division. This group would worry about trajectories to the Moon and back, including minding the calendar for those ideal lunar shadow lengths. They would roil through every conceivable way a mission could go wrong and how to then recover astronauts safely; they would sweat every drop of fuel, gulp of air, or charge of electric power needed during a Moon voyage. As engineer Aldo Bordano put it, "The first thing we did was design the mission. Then we had to teach the ground controllers and the astronauts what that mission was to look like and how to monitor it."[34] Aside from writing and teaching the script, the division also had to weave this expertise into the computer programs that would run in both the Mission Control Center and the spacecraft themselves. The planners describe nutty meetings of this era where they argued not only about solving problems but also over precious chalkboard real estate. They stood elbow to elbow in a cacophony of clicking, sliding chalk, marking task lists, differential equations, weights, and time points. "Problem was," my father said, "boards could be erased."

One of my father's first tasks for NASA was to go out and start hiring *other* new employees. The agency was growing rapidly, and only a handful of college programs even tried to prepare graduates for space work. Many new NASA employees attended days of impromptu classes held at Ellington Air Force Base, just to the west of the new Manned Spaceflight Center.

Engineer Lee Norbraten, for instance, accepted his NASA job offer right out of college, after graduating with a mathematics degree. "It was my first big adventure from home," he says. "I came to Texas without a car—this is not a very good thing." Once in Houston, he took a cab out to the new site and found what many new employees saw there: a sprawling, new campus with little else in sight. As my mother describes the 1964 scene, an old farm-to-market road provided the only route to the NASA center. "I don't think there was even a gas station out there at the time," she says. And there was definitely not a housing complex or dormitory.

Norbraten rented a small apartment many miles to the north, in the industrial town of Pasadena. Then he learned with dismay that he was too young to get a car loan. But he was a problem solver, and he started to circle his apartment building. "I went around the parking lot looking for a NASA sticker," he says. "I left notes offering to pay for gas if I could get a ride." He soon got a knock from a fellow who was glad to help. The man was from Arkansas, largely uneducated, and did maintenance work at the new space center. Their shared rides taught the young mathematician a few things. "It was my first encounter with country music." A popular tune of the day told the story of a tense double date: "You Can't Have Your Kate and Edith, Too." During those fifteen-mile rides to work, "I heard it more than once," Norbraten says.[35]

Engineer Aldo Bordano started at age twenty-two, right out of Texas A&M University. "I was so naive," he says now. "I remember the [application] said, 'Can you work at high places?'" So, he began to imagine himself working on the rocket support gantry next to the enormous Saturn V, "peeking in the window." An older coworker escorted Bordano on his maiden visit to the new center. "I was sitting shotgun in his car and I looked over to the right and there was a huge structure—it turns out it was a power plant. We drove by it kind of quickly and I looked over and I said, 'Is that where we're going to launch from?'" His coworker gave a casual nod in the affirmative.

"For another six or so weeks I thought that was the launch tower."

Bordano's early days give us a glimpse of the center's culture, as he soon was working sixty-five to seventy hours per week. One Saturday morning found him running and tweaking a computer program. He was so tired and so focused on keeping his stack of cards in order that he didn't see a little bottle on top of the chest-high IBM 1620. "I accidentally knocked that ink well over and it poured right down inside the computer." What would the ink do to the mysterious innards of this expensive device, a machine central to so many NASA projects? "I looked around and I picked that ink well up and I cleaned everything up and I looked at my run and it looked all right. Other people were coming in and I hung around kind of sheepishly. But their runs were working well, so I didn't ever have to tell anybody."[36]

The engineers arranged stacks of nightmares in obsessive detail. Eventually, for every possible Apollo mission that would come to be, engineers had planned and computed scores of failed and aborted missions. Some engineers specialized in the first eight minutes after a launch and each type of problem that could scuttle a mission before it got to space. If aborted, how would NASA get the astronauts back safely, and where would the astronauts land in the Atlantic? They asked and answered arrays of these questions for each conceivable problem. Others focused, for instance, on a return trip from the Moon to Earth. At what points in space could they safely adjust a mission's path and speed? When exactly would it have to leave the Moon to eventually land, during daylight, in the Pacific Ocean just west of Hawaii, near a fleet of waiting ships? They plotted a band of possible splashdown points and made sure the capsule wouldn't crash-land on any islands. If problems arose in any of the many orbits of the Moon, engineers computed the exact trajectories back to Earth from *each* of those distinct lunar laps. And still, surprises lurked.[37]

Max Faget's engineering division toiled just as busily on the new campus south of Houston. The testing facilities—Faget started calling them his "garage"—needed to test a spacecraft, its materials, and its design versus drastic temperature changes.

Seals around windows had to prove they would not leak when space was sucking on them with an unrelenting vacuum. Different modules had to prove that they could lock together during docking, at various speeds, with some forgiveness for the imperfections of human pilots. In 1964, he nervously watched the proposed Apollo spacecraft pack on the pounds, as idealized schematics became physical objects, with extra complications, extra materials, and extra subsystems added. Just some fifteen years after the scruffy kid who'd slept in his dad's car interviewed at Langley, Faget now employed 1,400 engineers. About three times that many also worked on the primary spacecraft via the contractor North American, based in Los Angeles.[38]

Faget also had a growing non-mechanical concern: preventing a deadly Moondemic. The National Academy of Sciences had just announced that "the introduction into Earth's biosphere of destructive alien organisms could be a disaster of enormous significance to mankind." Again, we might struggle to return our minds to this time, when so little was known of the Moon and what exactly we might find there. The risk of contaminating Earth with a Moon virus or a Moon weed were small, and many people scoffed at the science-fiction type of alarm, but just to be sure, Faget and a few of his engineers began to think about how they would quarantine and monitor returning lunar material. He became an early advocate for a lunar "receiving lab."[39]

My father never knew Faget well, but they intersected in meetings, and what he recalls squares with other accounts of the quirky genius. Many a meeting featured Faget, "glasses up on his forehead, eyes closed," my father says. "Everyone assumed he was asleep but he was wide awake." If someone in the meeting said something mistaken or ventured onto shaky engineering ground, Faget sat upright, eyes fluttering open, and issued a correction, sometimes in harsh terms.

Henry Pohl recalls some special cases of Faget feedback. "I remember I was making a briefing to him, and he told me, 'Now Henry, it's like a dark wool suit. [Your presentation] gives you a nice warm feeling, but nobody really notices it.'" Another time, Pohl was describing a new method for the small control thrusters, one that would save the Apollo spacecraft some valuable weight. "Now Henry," Faget said. "You're picking at cabbage."

"Well, I knew I [had] lost him," Pohl says. "Cabbage was plentiful and he was trying to tell me that [the idea] was not significant." He pauses. "It's a good thing, too." Pohl's idea, in this rare case, had been fundamentally flawed.[40]

In 1964, a major new residential development was underway in Clear Lake, but more immediate housing options were still limited for transplanted or newly hired engineers and their families. While a number of the Langley personnel had settled well west of the center, in Friendswood, Texas, others had settled to the south.

The Faget family opted for a home on Dickinson Bayou. "He basically designed the house himself," Max Faget's daughter Ann recalls. "With some unique features." For instance, "there was a long hallway to the bedrooms and at the end of the hallway, he had a little plug in the foundation, covered with carpet. But you could lift it up and there was a little [golf] hole so he could practice putting." His daughter Carol says their father built a bulkhead against the bayou in the late 1960s, to protect their yard from erosion. (The home is still in the Faget family, and while the bulkhead has held up beautifully for half a century, the floods of Hurricane Harvey overran the house in 2017.) Max had always loved being on the water. In the new home, he designed his own sailboats for local racing competitions, and he built his son, Guy, a little boat for exploring the bayou.[41]

My parents purchased a parcel of an old pecan orchard, about seven miles south of the new NASA center. Like the Fagets, they settled into a formerly working-class environment. The school buses mixed a few NASA kids with children of shrimp fishermen, mechanics, petrochemical technicians, and migrant field workers. In the years to come, most families of the space race inhabited the new developments around Clear Lake or other sprouting suburbs. Like Huntsville, the southern reaches of Houston experienced an abrupt cultural lurch—in terms of schooling and average income, as well as accents and traditions. But broader shifts were afoot.

The new president, Lyndon Johnson, accelerated an upbeat, progressive agenda. His Earthside projects included his Great Society programs. Meanwhile, he and Congress continued to increase NASA's budget, which surpassed five billion dollars for the year (over 4 percent of the total federal budget). The White House banked on the economy sustaining incredible growth, printing more and more money for ambitious projects.

To the president's way of thinking, certain societal changes were mandatory and overdue. Johnson signed the Civil Rights Act into law in early July, but the congressional delegations of the Deep South, where most of NASA now lived and worked, had strongly opposed it. Marlowe Cassetti, newly transplanted to Houston, recalls his new surrounds being even more segregated than those in Virginia. When he hosted a group from IBM including African American employees, he says, "We couldn't go out to lunch with them." With some relief, he learned that a local department store's luncheonette would accept an integrated group.

Meanwhile, the Student Nonviolent Coordinating Committee brought nearly one thousand young white volunteers to the Mississippi Freedom Summer to register black voters. In between the various engineering deadlines, NASA families would see TV footage and newspaper photos with waves of policemen beating and jailing volunteers and prospective voters alike.

Alabama's governor, George Wallace, remained an outspoken proponent of the old ways, and even as he proclaimed "segregation forever" throughout 1964, he found himself at odds with the more progressive-minded von Braun. During a meeting with the governor, von Braun and NASA director James Webb tried to defuse the tension. They politely asked of Wallace, wouldn't he want to someday visit the Moon? "Well," he replied, "you fellows might not bring me back."[42]

A legacy of the NASA versus Alabama disconnect persists in spirit today, where you can still hear a joke among the retired engineers. *You know what they say: the great thing about Huntsville is you're only fifty miles from Alabama in any direction.* The space centers, with their long hours and stacks of impossible problems, along with federal leadership and mandates, could function as islands. For the most part, the engineers didn't have spare time or energy to consider careening social changes outside their labs and offices. At times, they didn't even think much about their families or about Earth in general, until an event could yank them back.

One Marshall Center engineer described joining von Braun's workforce and moving to Huntsville in 1963; he was new to the South. With his first real salary, he bought a car, an Impala. One warm September Sunday, he was living a cherished American dream of washing an automobile in the driveway and listening to the radio, when the announcers cut in with horrifying news: The bombing and burning of a black church in Birmingham, one hundred miles to the south, had taken the lives of four young girls. The white engineer stopped fussing with his car and stood stunned, as the white disk jockey transitioned directly to music in a jovial tone. An evil-sounding, mocking laugh opened a popular surfing track before intoning, "Wipe Out!" The story still gives him chills.[43]

By all accounts, the racial animosity and injustice genuinely moved and disturbed Wernher von Braun. Coworkers warned von Braun about public stances. They told him he would get a cross burned in his family's yard if he kept speaking against racism. But von Braun described his arrival in America and the questions he had faced. "Where were you? What did you do? . . . What did you do when people were disappearing?" he said. "That's not going to happen again. . . . I am not going to sit quiet on a major issue like segregation." And he didn't. He spoke for desegregated schools and voting rights. In December of 1964, he addressed the Huntsville Chamber of Commerce: "I think we should all admit this fact: Alabama's image is marred by civil rights incidents and statements." At another event, he told an audience that "all these regulatory barriers form a Berlin Wall around the ballot box."[44]

Despite his good intentions and some sincere efforts, von Braun made little progress in integrating his Huntsville workforce. Throughout the 1960s, while 18 percent of Huntsville's population was African American, the Marshall Space Flight Center featured less than 1 percent African American employees. Segregated schools in Alabama, including segregated higher education, meant few African American students received the engineering training needed for rocketry. And von Braun's recruiters found it nearly impossible to entice any non-white person from outside the state to Alabama, given its overwhelming association with the words of Governor Wallace.[45]

Though any investigation of von Braun's life in America finds fawning stories of a kind man, a great listener, and a natural leader, many people find it difficult to reconcile his stance in the Civil Rights era with his participatory Nazi past. The same man who spoke truth to the power of Wallace, in some cases questioning him directly, had accepted promotions as an SS officer, and he had witnessed the gruesome underground facility, Mittelwerk, where slave laborers died en masse building V-2 rockets to his specifications. Records show that von Braun even sifted through concentration camps, looking for engineering talent. But the 1944 and 1964 versions of von Braun are not necessarily at odds. Any gesture against the Nazis certainly would have (and arguably did) put his life at risk, whereas a strong stand in Alabama did not. As with the story from his co-workers, he may have had a sincere change of heart, the emergence of a strong moral gyroscope. People are allowed to grow between their youth and their middle years. But we could also cynically say that in each case (Nazi Germany and 1960s America) von Braun recognized the hand that fueled his rockets. George Wallace wasn't paying the bills. Meanwhile, the federal government and NASA leadership had told him in 1963 to start a serious affirmative action program in Huntsville. He was truly following his orders again.

The day after von Braun addressed the Huntsville business community, chiding Alabama for its self-inflicted harm, NASA prepared a significant test launch in Florida. The region enjoyed one of its more gorgeous December days (calm winds, sunshine, and 68° Fahrenheit), buoying the hopes of the assembled. Engineers needed to test an intermediate spacecraft called Gemini, a two-seater model that would bridge the earlier Mercury capsule to the eventual Apollo spacecraft. On this day, they would use a Titan rocket to lift the empty spacecraft briefly into space. They needed to test its performance in space and its newly designed heat shield on return. The Titan used the nasty hypergolic fuels that burned on contact with one another, no spark required. As the countdown reached zero, the engines roared but then went quiet, as an acrid red cloud of fuel billowed around the launch pad and rocket. A simple plumbing problem had caused this misfire. Engineers, trying to save weight, had simply shaved one of the pipes to be just a little too thin to withstand launch. After the unnatural caustic smog cleared, engineers rubbed their temples, gathered themselves, and got back to work.[46]

8

1965—SATURN BREATHES

Henry Pohl's team ran around-the-clock tests for two weeks at a time working to perfect Apollo's forty-four control thrusters, each the size of a hand. They had to be sure the thrusters would reliably cycle on and off, over and over, during longer missions. Engineers watched an orange residue encrust the inner test chamber walls. This leftover, congealed from thruster exhaust, was still explosive. "We could put some on an anvil and tap it with a hammer and it would pop like those old cap pistols," Henry says. They sent some to a standards technician in Boulder, Colorado, to assess the material. "The next day he called back and said it was 1.6 times stronger than TNT." Soon after, late one night, a frightened technician ran into the laboratory. He said that the residue now coated the parking lot outside; the system must have a substantial leak, and they were spreading an explosive powder across the center. "We ran out and with cigarette papers scraped [a sample of] it off the cars," Henry says. His boss wanted to shut down the test facility immediately, but Henry couldn't believe they had such a bad leak and he wanted to analyze what was falling outside. He took a bit of the powder to a local expert. "That University of Houston chemist looked in the microscope for a very long time, stood up and said, 'If I didn't know better I would say it's pollen.' I snapped my fingers and said something like, 'Son of a gun.' . . . [T]hat area [around the parking lot] was solid goldenrod in full bloom."[1] Though the engineers could exhale for now, the thrusters would bedevil them for months to come.

Wernher von Braun and his Huntsville team worked on the larger scale propulsion, as they prepared to test their behemoth: the first stage of the Saturn V rocket, where five of the world's most powerful engines would light in concert. The Saturn V consumed roughly one-third of the total Apollo budget throughout the 1960s, and about half of every dollar spent on the Saturn V rocket went to testing.[2]

The enormous test stand that had sparked a Huntsville labor strike in 1962 was ready. In total, with its upper crane, the test stand stood over four hundred feet high. Technicians bolted the rocket's bottom stage upright, suspended some one hundred yards above the earth. This temple of iron and cement began deep underground, reaching forty-five feet into the red soil until it anchored to bedrock. Its deep grip aimed to keep the Saturn engines from pulling the test stand from the ground.[3]

Given the stand's non-public purpose, engineers wasted no time on cosmetics. It looked (and still looks) like an ideal set for a post-apocalyptic B movie: metal bolted to concrete bolted to metal. Sturdy but built in a hurry, some segments were not perfectly level. Bolts securing a rectangular metal plate, for instance, sat different distances from their respective corners. The stand only had to strap in the Saturn's first stage like some baby in a high chair, to be fed swimming pools of kerosene and liquid oxygen.

From the upper levels of the stand, an engineer could view a half dozen other test facilities interrupting the forests and fields. One of the closest was the "dynamic" test stand, a black monolith-like structure, fully enclosed. Its name may suggest a moving building, but here the engineers would assemble a Saturn rocket and then vibrate it, poking and prodding the rocket with various timings to understand exactly how its various parts would rattle and shudder in the violent launches to come.

The static test stand featured a gigantic scoop to corral fiery exhaust. The scoop's semicircular curve, ribbed and segmented, resembled a giant earthworm lifting itself from the ground. Each segment was plumbed for cascades of water during a test firing, to prevent the test stand from melting and to dampen the tremendous roar (see Figure 8.1).

At a basic level, the engineers weren't sure what to expect. When firing together, the five F-1 engines would become the most powerful human-made device outside of an atomic weapon. Engineers weren't sure if nearby structures could survive the first test, so they ordered staff to vacate every building anywhere near it, except for the blockhouse.[4] About three stories tall and a football field away from the test stand, the blockhouse hunkered within reinforced concrete walls at least a foot thick. (Like the test stand, it remains there today, and it's not clear what could ever destroy it.) Square windows, each one foot by one foot by one inch thick, decorated the side facing the test stand. The edges of the building were rounded, offering no purchase for the possible winds of an explosion.

The initial test of the first stage, in the spring of 1965, ran for just over six successful seconds. All five F-1 engines lit together, burned without trouble, and fell silent together. Engineers who witnessed the test described it as a repeating shotgun blast, with sound waves of such strength that their shirts would flap violently against their chests. All meetings around Marshall stopped with the incredible noise of the test firing, and not by choice. One engineer, on an important phone call far away from the test stand, crawled under his desk and yelled into the receiver, not sure if he was still connected. Another had taken his family shopping in Birmingham, some one hundred miles to the south. As they walked in the J. C. Penney department store, all the windows around them began to rattle. The engineer's wife asked him what was happening, and he calmly said that Marshall was probably testing the Saturn V.

FIGURE 8.1 The static test stand's curved flame bucket at the Marshall Space Flight Center. To help emphasize the facility's scale, current Test Laboratory Director Ralph Carruth stands near the bottom. (Photograph by author.)

Indeed, Birmingham registered an earthquake-like event, but it arrived through the air, from sound waves of too low a pitch for the human ear to register (see Figure 8.2).[5]

The tests became a spectator sport for some. "There were two or three of us that would call our wives," engineer Bob Austin says. "And there was a place to watch from a safe distance." They would call about thirty minutes before the test would start, but the tests themselves were usually brief. "I'd tell my wife, don't get caught looking the other way . . . watch the pump house, and when the black smoke comes out, that's when we're about to fire." The pump house kept a row of monstrous diesel engines, and right before a test firing they forced torrents of water toward the stand to coat the inside of the blast scoop. A tall person could easily stand up in the pipe

FIGURE 8.2 A static test firing of a single F-1 engine in 1965. With five of these engines firing together, the Saturn V's first stage consumed fifteen tons of fuel per second. (NASA photograph.)

between these pumps and the stand, and the pumps could move a million gallons in just a few seconds.[6]

While successful tests of the first stage encouraged von Braun, he received dire news on the *second* stage, as a version crumpled and failed during a structural strength test in California. The second stage of the Saturn V used the more advanced hydrogen engines, the ones requiring wholly new welding techniques for their super-cold fuel tanks. The second stage was so far behind schedule that NASA eliminated the plan for a series of test models and asked the contractors to focus on the version they would use for the actual missions. One of von Braun's deputies warned him that the project was "out of control" and could, on its own, push the entire Apollo program past their deadline.[7]

If the Saturn's thick breath came in halting bursts, the more rarified issue of human breathing confronted engineers in Houston. Astronauts would need oxygen, with their lungs accustomed to our thick atmosphere near Earth's surface. But to have a full atmosphere's worth of pressure in a spacecraft is to increase the stress on all its joints, welds, and seals. NASA opted for a decreased pressure for its various space-craft. They aimed for roughly one-third of normal air pressure (about that found atop Mount Everest). It was much easier to have things leak-proof with the lower air pressure inside; the eager vacuum of space sucked on each seam with one-third the force that it would have otherwise. But that led immediately to a second problem: humans usually breathe thick air that is about 20 percent oxygen and 80 percent nitrogen. Thinner air, of course, offers less oxygen, and the brain starts misfiring a bit. With low enough oxygen levels—never mind making a few bad decisions—an astronaut could die.

Increasing the percentage of oxygen in every gulp of spacecraft air offered one solution, but Max Faget and others argued against a mix early on. Having two gases (nitrogen and oxygen) meant twice as much piping and twice as many systems that could fail during a mission. The simplest solution involved a 100-percent oxygen cabin atmosphere, at a lower pressure. The astronauts would have all the oxygen they needed, but the pressure in the cabin could be fairly low before the astronauts were in danger of asphyxiating. For an astronaut donning a suit and floating in space or walking on the Moon, engineers wanted to decrease the pressure even further, to about one-*fifth* that of our normal atmosphere. At that point, having *any* nitrogen risked a nasty condition called the "bends" where a body would start sucking nitrogen into the blood stream. This could strike an astronaut, much like a scuba diver, without warning, with a variety of dangerous symptoms ranging from intense joint pain to paralysis.[8]

A perilous lesson in pressurized space flight occurred during a 1965 Soviet mission, but at the time NASA only heard the great and glorious result announced: a Russian named Alexei Leonov had become the first human to float freely in space, without the protection of a spaceship. He floated at the end of a tether, supremely vulnerable yet feeling peaceful, until his colleague reeled him back toward the ship. Only decades later did America learn the whole story.

Before drifting into space, Leonov had lowered the pressure of his suit to roughly one-third the normal atmospheric pressure, and while floating, he was just fine and even thrilled for fifteen minutes. But then he sensed a dire problem. "I realized that my feet had pulled out of my shoes and my hands had pulled away from my gloves," he later said. "My entire suit stretched so much that my hands and feet appeared to shrink." Unlike during various tests on Earth, the suit in space had come to resemble the Stay Puft Marshmallow Man. At every point, it strained outward like a rigid balloon. Leonov's gloves no longer fit, and he didn't have the strength to even bend the suit fingers. "It's like bending a pipe," NASA engineer Larry Bell said later. "Gloves are still a challenge and always will be." And as Leonov tried to clamber back to safety, the bloated suit no longer fit through the hatch to the Voshkod capsule. He tried to bend at the waist, but the suit wouldn't yield. His pulse and blood pressure shot up and he began sweating mightily as he fought his rigid carapace. To save himself, he violated mission orders, deflated his suit to an even lower pressure, and squeezed inside the craft. Even after re-pressurizing the capsule, the mission did not let up on poor Leonov and his fellow cosmonaut. Their automated navigation system failed on their return to Earth, and they fell in the Ural Mountains, some two thousand miles away from their anticipated target. Suddenly, the men had descended from the space age into a traditionally grim Russian short story. Soon, hungry wolves surrounded their capsule, far from any human village. The cosmonauts shivered through the night and only heard the voices of rescuers the next day. As usual, outside of a handful of Soviet engineers, the rest of the world only heard of a cosmonaut's happy, carefree spacewalk, another bragging right for the program.[9]

In America, spacesuit activity reached a kind of Olympic pinnacle in 1965, as NASA hosted an Apollo competition between three rival manufacturers, including the decidedly non-space-age International Latex Corporation. A veritable track meet, the competition included twenty-two diverse events. How mobile were humans in the suits? How did the suits handle extreme temperatures? Pressures? How exhausted would an astronaut be trying to move within the suit? How easily could an astronaut fold a suit when it wasn't needed? How wide were the shoulders?

The helmet popped off one of the suits when it was pressurized. A different suit got stuck in the entrance to an Apollo mock-up craft. At the end of the day, the supple, multi-layered, seamstress-powered suit from the Latex Corporation won twelve of the twenty-two events. NASA officials ranked them in first place, and according to their official report, "there [was] no second place." The other suits were disqualified.[10]

Meanwhile, NASA enjoyed a rare media victory and a spacesuit success as the first American astronaut floated in space that summer.[i] Much less secretive than the Soviet program, NASA shared live audio of the event, including astronaut voices, with the world. (They did not yet have television cameras in the missions.) The astronaut's umbilical cord included a little tube for oxygen, a few electrical wires, and a steel cable. Everything held together, and the astronaut described a kind of euphoria. Not needing to move in any specific way, he did not encounter the problems of Leonov.

On Earth, activity in Houston, Huntsville, and Cape Canaveral reached an absurd pace. Test facilities came online, with varying success, and systems in every case needed debugging.

One especially ambitious laboratory aimed to create the vacuum and the wild temperatures of space. It's one thing for scientists to create a Moon-like vacuum within a metallic chamber the size of a kitchen oven—it's possible but takes weeks of fussing with clamps, cleaning every fingerprint or grease smudge from the chamber, and then patiently finding and fixing the leaks. But NASA's new environmental chamber held the volume of a gymnasium: a cylindrical chamber over one hundred feet tall and about two hundred feet in circumference. To move spaceships in and

---

[i] Astronaut Ed White, mission Gemini 4, spacewalk of June 3, 1965.

out, the chamber featured a front door about forty feet across. When closed, it had to become leak tight. Not only had humankind never built a vacuum chamber so large, with no air leaking in, but engineers also wanted the room to feel the chill of space. The walls of the chamber contained jackets of liquefied helium, just a few degrees above absolute zero. After several years of careful design and construction, the engineers felt confident enough to give the chamber a try in the spring of 1965, and they gradually pumped the air out of it. But well before achieving the vacuum of space, the large door crumpled inward, essentially imploding.[11]

Nearby, one of Max Faget's busy workshops stuck to basics. "We had this damned plywood mock-up [of the Apollo's main command module] sitting there in the shop, and it had a door that would open and a mattress on the floor," said Faget's longtime collaborator Caldwell Johnson. They were testing practical matters. For instance, could an astronaut actually crawl out of this Apollo craft if need be? A volunteer tried climbing from the cone-shaped wooden model of a space ship, while engineers took notes and timed him. "Well, it was made out of three-quarter-inch plywood, and you know how the edges were," said Johnson. "And that guy was blood from bottom to top at times. . . . [H]e'd banged against the side and missed the mattress and done all these things and skinned up [both] arms. They'd say, 'Just one more time, John, and let's try it this way.'"[12]

But Apollo wasn't the only game in town. The space program's second phase, a kind of interstitial program to work out the kinks of spaceflight, shifted into full speed, aiming for a mission every other month. The engineers express a surprising range of feelings on Gemini. Some detractors believed it stretched NASA too thin or was simply a ploy to stay in the headlines while Apollo preparations progressed behind the scenes. Others saw it as absolutely necessary, teaching an adolescent agency critical lessons on life in space.

Marlowe Cassetti found himself straddling two programs at once. In his first years in Houston, he'd gone from a man in his late twenties managing six older engineers to the leader of about thirty-five, including at least a few new hires mercifully younger than him. Now NASA charged him with both helping guide Gemini launches from the Cape and leading a team in planning trajectories for Apollo test missions. As he routinely abandoned his Houston team to attend launches at the Cape, he had a difficult time keeping both programs happy. "I was getting a lot of heat," he says, "because I was spending so much time on Gemini." Some engineers didn't try to hide their competitive feelings. Cassetti was initially surprised to hear a Gemini colleague claim, "We're going to beat Apollo to the Moon." He describes a "fall-back position" that his colleagues wanted to present to NASA leadership: If Apollo kept falling behind schedule, their two-seater Gemini craft could find a way to take a spin around the Moon, even if it couldn't land.[13]

One of Gemini's more accepted goals involved testing a human's ability to stay in space for increasingly long stretches. In 1965, NASA finally felt comfortable enough to move forward with longer manned missions in orbit thanks to a more obscure project called Pegasus.

Apart from whether the human body could withstand so much time away from Earth, scientists had another significant worry. Small, undetectable space pebbles, moving tens of thousands of miles per hour, could potentially perforate the hull of NASA's spaceships. The transplanted German physicist Ernst Stuhlinger recalled that "von Braun told me one day we should know more about the meteorites in orbit. . . . Do we have to expect hits and damage? How many are there, and how big are they?" Stuhlinger's small team had set out to find answers and devised the Pegasus satellite. "Pegasus was a winged system," he said. "And these wings were only sensors for meteorites." (See Figure 8.3.)

Stuhlinger had worked with von Braun from the early V-2 days and provided a leading light for pure space science in Huntsville. Beyond his many contributions to Apollo and several unmanned programs, he toured the country to talk with astronomers about the idea of observing the universe from the open clarity of orbit, laying the groundwork for the eventual Hubble Space Telescope.

In Pegasus, Stuhlinger's team created a micrometeor detector of beautiful simplicity. The wings were sandwiches: two sheets of thin metal separated by a non-metallic layer that could not pass electricity. When struck by a speeding space pebble,

FIGURE 8.3 Personnel test the Pegasus micrometeoroid detector array, which unfurled to a width of ninety-six feet in space. (NASA photograph.)

the incredible momentary heat briefly created a pocket of hot gaseous leftovers ame-nable to electric current.[ii] For less than one-millionth of a second, the two layers of metal could communicate, and the satellite would record this blip. Once the gas quickly dispersed, the two sheets of metal went back to chilly separation, awaiting the next impact. The size of the electrical discharge in each impact directly described the size of the meteor.

The large, unfolded wings of Pegasus, with more than fifty times the profile of an Apollo mission, recorded only seventy impacts in three months, and the vast ma-jority were not of a size that would threaten Apollo. By 1965, NASA could cross the hurtling-space-gravel worry off its list, and Gemini moved full steam ahead.[14]

They decided that the fifth Gemini flight would aim for an unprecedented *week* in space. Engineers could test an astronaut's ability to endure the time needed for a full Moon mission, but America could also gain bragging rights over the Soviets in the space duration category. In the end, this mission demonstrated that all of NASA had significant work to do, with only half a decade to go. First, the launch experienced a version of the "pogo" stuttering so severe that the astronauts briefly lost their vision. Once in orbit, the mission tested a newer technology: Fuel cells combined oxygen and hydrogen, liberating energy *and* creating clean water as a byproduct. But on this trip, the fuel cells hummed along too efficiently and the astronauts had more water than they could drink. Soon, they started filling plastic bags with all the extra water, which littered the spacecraft's cabin, floating about like so many carnival bags missing their goldfish.

Next, the ship's planned maneuvers in orbit became impossible, as first one and then other little thrusters malfunctioned. Instead of running through a series of tests to alter their orbit, the astronauts had to float idly in space, letting their ship just drift and slowly tumble along its original path.

Even the mission planners and data analysts contributed problems. When coming in for landing, the Gemini capsule badly missed its intended ocean target, splashing down well short of the recovery ships. Why? A clutch of mortified engineers confessed: In calculating the ship's path back to Earth, they had neglected to account for Earth's continued spinning as the capsule descended.[15]

But "Gemini V" was far from a disaster, and America finally held a temporary space record: the longest manned mission, at nearly eight days. NASA recovered their astronauts, and doctors vented a long sigh of relief when the astronauts' bodies still functioned. Some had wondered if so many hours in weightless orbit would be too much for a body evolved for the constant pull of gravity; they even worried that the astronauts might collapse and die upon their return. Doctors had already noted that

---

[ii] The remnant puff was technically a plasma.

astronauts came back to Earth with higher heart rates and lower blood pressure. So, they were ready to give CPR to astronauts in the rescue rafts that plucked them from their capsule. None of that was necessary. The astronauts were a little wobbly and a little smelly but otherwise fine.[16]

At the same time, a key NASA engineer and manager graced the cover of *Time* magazine in a dramatically painted portrait. Christopher Kraft had become a minor NASA celebrity, especially rare for a non-astronaut. He and his team in the Mission Control center were now familiar to Americans. Kraft had been part of the program since the beginning at Langley, and in the earliest days, he had led the dreaming, design, and eventual function of the control center. America seemed to eat it up: the banks of lights, phones, and monitors; white-shirted engineers absorbing data, finding patterns, and debugging the unexpected; the key decisions made and distributed in a complex, ever-changing environment.

The Austrian Henry Koerner painted Kraft's face in wide strokes and many shades of beige. In one sense, the portrait actually spoke to a deep truth of NASA at the time: homogeneity. Most of agency looked a whole lot like Kraft. From Faget and von Braun to Pohl, Cassetti, and Brown, NASA had grown quickly from deep 1950s roots as masculine as they were white.

Those hiring and filling the ranks probably had little intent to keep replicating themselves. The educational pipeline for young scientists, mathematicians, and engineers had filtered the applicants before they showed up to interviews. For women and people of color, technical education included extra hurdles beyond equations and circuits.

Later in the Apollo era, my father hired a skilled young engineer named Cynthia Wells. But in 1965, she was surviving life in a large lecture hall of fellow students. "I was the only female in the engineering majors," she says. In an advanced calculus class, her professor often addressed her directly. "They think the females are less bright. He would ask, 'What don't you understand today?'" The professor calibrated his course by measuring her: Surely if she understood it, all the men had mastered it as well. Or, more charitably, perhaps he also noted a fearless confidence in Wells. In any case, his routine became something of a joke among the students. Wells aced her classes, and in the moments before that calculus class, desperate boys would beg her to ask their questions.

When she eventually started her twenty-year run at NASA, she felt "protected" by her male colleagues and even enjoyed the added attention of being the only woman in many of her meetings. She recalls being deployed strategically for certain negotiations, where rival engineers—from Huntsville, for instance—would grill her less harshly than they would have a male colleague. In the end, was she one of the guys? Not exactly. "My husband did stuff with them," she says, speaking of softball games and after-work bottles of Lone Star beer. "I wouldn't have enjoyed that stuff anyway."[17]

Many pioneering female and minority engineers of this generation emphasize the positive at NASA. Retired engineer Wesley Ratcliff says he definitely benefited from 1964's Civil Rights Act. "Being African American and a male," he says, "I had a lot of companies that were in interested in me." A 1965 interview at NASA impressed him; he spoke at length with three very different groups, and he was especially drawn to the idea of rendezvous, directing one spaceship to grab another while damaging neither.

Born and raised in a poor town in the piney woods north of Houston, he recalls the generosity of a white rancher who'd hired him during the summers, helping Ratcliff earn a little money. And he still appreciates the efforts of his high school physics teacher in their segregated school. When Sputnik went up, the teacher set aside his lesson plans and taught the students how the satellite worked, and when America later got Explorer 1 into orbit, the teacher spent another class going over the differences between the two satellites. This kept a keen boy's mind simmering.

Ratcliff's friends and family encouraged him to go into medicine in college. "But I didn't get past the frogs," he says. "I didn't care what was in there next to what." He opted for something different. "In physics, I didn't have to memorize what went where. I could prove things. Could move this equation over here, and boom—it works out." He earned a physics degree from Prairie View A&M, a historically black college, and as an ROTC student, he followed graduation with four years in the army. He emerged in 1965—just in time for the new affirmative action policies—and he accepted the offer from NASA.

With no complaints, Ratcliff straightforwardly describes the bias he encountered. Many engineers assumed he couldn't do the work. "You had to prove yourself, and that's what I did, frequently." He fondly recalls one colleague who gave him the benefit of the doubt from his first day forward. "Immediately," he says, emphasizing each syllable. "That makes a big difference." As an employee in the math-heavy mission planning efforts, he once helped a fellow engineer correct some orbital calculations on a chalkboard—they worked for most of an entire day together. "Where did you learn that?" his white colleague asked.

"I went to A&M too," Ratcliff replied. "Just not your A&M." In the retelling, he makes sure to add that "we went on to be good friends."

Ratcliff stayed on through the bulk of the Apollo program and now credits NASA for his own growth. "Those years gave me—how do I want to say it . . . the knowledge that I wasn't as stupid and dumb as I thought I was." He faced the bias and the occasional, offensive "vernacular" but the organization focused on solutions, and in what he calls his "small way," he owned and solved a series of problems.[18]

While Ratcliff, my father, and others busily computed trajectories, orbits, safety parameters, and fuel supplies, others in Houston tried to simulate two colliding spacecraft that could somehow dock with one another. Since Phase II of manned spaceflight sought to prove that a rendezvous could work in space, engineers built a docking dress rehearsal that greatly improved on the ice-rink version. They designed a full-scale simulation, with all possible motions and rotations, for the lunar lander and the main Apollo spaceship, the ones that would need to come apart and then dock again while orbiting the Moon. NASA contracted a machine company from Pennsylvania. "They were noted for bowling alleys, for having the equipment to set the pins," engineer Tom Moser recalls. "They had that kind of robotic capability." Instead of having full models of the lander and the Apollo module, this replica would just have the pieces from each ship that would have to mate for successful docking—one would have a "probe," the more male part by analogy, and the other had a "drogue," the more female part. Computers would eventually control both of the dancing pieces, running various speeds and angles to see when exactly the probe and the drogue could come together and mate versus when they would bounce away, thwarted by geometry or physics.

Engineers eventually had the parts moving through the large vacuum chamber, using a choreographed ballet of hydraulic pistons and servo motors to practice the motions of ships circling the Moon. These trials included the temperatures and pressures of outer space. But the intense vacuum introduced new problems. Most materials in low pressures tend to "outgas," or slowly release loose atoms. As materials outgas, they essentially ruin the vacuum, since the chamber always has an "atmosphere" of these pollutants seeping from the materials. (It's a common problem for any scientist using a vacuum system.) Stainless steel offers one of the few surfaces that does not outgas much. "The test device—it was huge—it was mounted on a stainless-steel plate," Moser says. "It was the largest stainless steel plate ever produced because you put the whole thing in the vacuum chamber." How big? "Five inches thick, and fifteen feet by ten to twelve feet wide probably."[19]

To prove this joining could work in space, however, NASA had an even larger problem. Could they really persuade two ships, launched separately and moving at tens of thousands of miles per hour, to achieve nearly the same orbit and then to have them gently approach one another? Rendezvous requires three things.[iii] First, the two ships must maneuver to be perfectly "coplanar," or orbiting within the same mathematical plane. Picture cutting a globe in half and inserting a wide, stiff piece of paper between the two halves. A satellite orbiting Earth will move in a loop that you could draw on such a sheet (or plane). But we could have chosen an infinite number of ways

---

[iii] With thanks to Frank Hughes and Ken Young for the detailed rendezvous conversations.

to make that original slice of the globe. The paper could have bisected the North and South Poles, or it could have cut through the equator, or it could have come through at any particular angle, like one including Hawaii in the Pacific Ocean and Botswana in southern Africa.

Once the two spacecraft inhabit the same piece of paper (a feat that can consume a *lot* of fuel), they next must inhabit two orbits that are similar but not quite identical. Picture an ellipse or circle. Now draw a smaller one that is precisely the same shape, so that the two orbits always have the same distance between them, at every point. Putting the ships in *different* orbits might seem counterintuitive, but this is the only way that one ship can catch another around a planet. The ship on the smaller orbit needs to start *behind* the one in the larger orbit, because the closer orbits have greater speeds (sort of like the inside versus outside of an oval racetrack). For NASA's rendezvous attempts, the orbits might be just ten miles apart; at that distance, the faster ship is catching up at the gentle speed of twenty feet every second.

Finally, as the faster ship catches its slower and higher partner, it maneuvers carefully upward to finally join the higher orbit. Now the ships are incredibly close, in the same orbit, moving at nearly the same speed. A few last maneuvers—like those practiced by Moser and company in the vacuum chamber—bring the probe and drogue to a locked embrace. These last, tiny steps can still be vexingly difficult. Moving in circles renders our natural human intuition to move in straight lines counterproductive, and success was elusive for astronauts in the first rendezvous attempts, frustrating exercises that deserved a "Yakety Sax" soundtrack.

Engineer Cathy Osgood (she who composed the pro-and-con NASA decision list with her husband) recalls the earliest times of computing rendezvous. At first, they planned a simpler approach. They would launch the first ship into an orbit, and then "just launch directly into the orbit and rendezvous." But, "Oh no . . . if your launch [of the second ship] is delayed [even a few minutes], your plan has gone." She and her colleagues next considered using two naturally different orbits that would intersect in at least one point over Earth. But calculations showed that the ships would be hurtling toward one another at dangerously high speeds. The engineers eventually arrived at the recipe outlined previously, involving an inside track and an outside track.[20]

Even though this method eventually worked, they encountered further difficulties. When two craft are in different orbits around Earth, they each encounter a slightly different amount of residual atmosphere. The "lower," faster one fights against more drag from air resistance. This proved an elusive effect to master in engineers' predictive calculations. Furthermore, to be precise, we can't treat Earth as a perfect sphere. It bulges around the equator, and its gravitational tug varies from one orbital location to the next—not by much, but enough to cause complications. Engineers had to treat many of these factors as mysteries, running hundreds and thousands of computer simulations to understand the whole range of possibilities for a given mission.

All the while, a Gemini Earth-orbit mission in 1965 would still have some gaps of radio silence more than ten minutes at a time. NASA had worked its way to an unprecedented global communications network, with some twenty-five worldwide communication stations. When they'd had to evacuate the Zanzibar station in the midst of a coup, they set up another in nearby Madagascar. NASA built a station on barren little Ascension Island in the south Atlantic, despite conditions so brutal they forbade families from visiting the engineers stationed there. The network even included, during missions, a patient navy vessel bobbing in the Pacific. Despite all the progress, NASA had to accept that any capsule would spend time in dead zones in between their stations. During the more stressful and troubled missions, astronauts would enter the silent times alone, and ground-based engineers could only run their calculations and fret.

Worry and broken communication mirrored what some of the engineers' households experienced in this era. "The bad part of that is a lot of families suffered," says engineer Ken Young, echoing the confessions of many. "Especially Gemini, I worked six and a half days a week. I'd go home at noon on Sunday to watch the Oilers play . . . and the family suffered. That was our fault." Divorces became commonplace in the engineering ranks.[21]

Rushing into the autumn of 1965, the schedule at Cape Canaveral reached a sort of steady-state crisis. Engineers endured hectic launch days crowding ever closer on the calendar. Launch days could start at two in the morning and, even if all went well, end at midnight, after refining the schedules for the next launch and starting to refurbish the launch pads. If one task or mission fell slightly behind, the team juggled the densely woven schedule to maximize the progress of every hour.[22]

A striking example of this shuffling came in the last months of 1965, as NASA tried to achieve one or both of two milestones: rendezvous of two spacecraft (getting close to one another); and an actual docking of two spacecraft (linking them together, probe to drogue). In late October, NASA had planned to have a manned capsule launch right after an unmanned launch. The manned capsule was to chase an empty, dormant segment of the unmanned rocket in orbit and briefly attach to it. However, as the astronauts waited on the launch pad, the first rocket (luckily with no astronauts on board) fractured to pieces as it ascended. With no target for docking, NASA canceled the second launch.

The planners and engineers huddled: before Christmas, they could try something else. They quickly scheduled two manned missions, hoping two manned capsules could attempt the first rendezvous before the year's end.

The first mission went up without incident on December 4th, and the capsule found itself in patient orbit, awaiting its twin's launch about a week later. The waiting was not entirely boring, unfortunately. Fuel cells, energy-supplying innovations that they were, misbehaved again, each time triggering alarm bells and flashing lights within the Gemini capsule. Imagine dealing with a significant fire alarm while strapped into a closet. As one astronaut later said, it was like "eating, sleeping, working, and going to the bathroom stuffed into the front seat of a sports car."[23] As the fuel cells kept them awake, the astronauts grew increasingly tired and cranky.

The second launch nearly ended in tragedy. The Titan rocket, delayed from October, ignited on December 12 with, as engineer Christopher Kraft called it, "the customary belch of gray and red smoke," but then the sound stopped, with the rocket having barely shifted in its seat. An anxious pair of humans sat on top of an enormous explosion waiting to happen, as two poisonous and volatile chemicals longed to combine beneath them. In this case, calm human intuition saved the day. By the book, the astronauts were to pull an ejector ring and have the capsule fling them harshly out over the launch complex. But they decided to ignore their training because, in the quiet, with no sign of an explosion or fire, it seemed safer to sit tight. Fuel pressures eventually ebbed, and NASA safely removed the astronauts from the rocket. In the end, engineers found two little problems: an umbilical cord had disconnected from the rocket a tad early, leading the rocket to assume a major malfunction. Next, engineers found a little plastic dust cover in a fuel line; someone had simply forgotten to remove it when putting the rocket together before launch.[24] But after some round-the-clock work, they sent the rocket on its way again just three days later.

Now, in the last weeks of 1965, NASA had two manned spacecraft trying to find one another in the immensity of space above Earth. Soon, the mathematics of rendezvous provided a miracle. One capsule caught sight of its twin, first appearing as a bright star, with the inner track slowly catching the outer track. The ships grew in one another's windows until the two pairs of astronauts were just 130 feet apart, each traveling at absurd speeds above Earth. Rendezvous had worked, and several engineers have marked this moment as the pinnacle of their NASA careers, after years of proving, on paper, that it could happen. By gently using their thrusters, astronauts got their ships so close that they could have opened a hatch and literally touched one another. Fresh-faced and energetic astronauts peered out from the second capsule, seeing more scraggly and exhausted faces in the first. The first capsule had already been in orbit for a week and a half. Food crumbs, dried skin flakes, and droplets of pesky, escaped urine floated around the astronauts.

The second ship, with its happier campers, soon made its way back to Earth, while the first capsule slogged along to complete its long-duration mission, testing how long humans could stay in space. By the end, with their fuel cells dead and most of their electronics switched off to conserve power, one of the astronauts said spending four-teen days in the capsule was like "living in a men's room"—and not a large one. Even removing a spacesuit was a major geometrical puzzle. One astronaut, out of tasks, just read *Roughing It* by Mark Twain, dreaming of open skies and elbow room. When they at last splashed down, navy divers swam to their capsule to recover the 13.75-day space veterans. Three frogmen attached a floatation collar to keep the capsule upright, as usual, and then helped the crew open the hatch. In this case, despite the rigorous training and seaworthiness of the divers, two of them apparently wretched in the ocean after getting a whiff of the cabin air and the fusty astronauts within.[25]

By the end of 1965, some of the blush had arguably faded from the spacefaring rose. And while plenty of progress could be found, troubles multiplied just as quickly. Engineers scratched their heads over how they could hold a massive Saturn V rocket upright before its launch. They designed a new launch tower with nine "swing arms," each weighing around twenty tons and wide enough for a lane of car traffic. In ad-dition to holding the rocket upright, the arms would convey fuels, oxygen, and elec-trical power. And they had to be nimble, retreating in two coordinated waves from the Saturn V rocket at just the right moments of a launch. The arms, while critical to Apollo, aimed to grip a not-yet-finalized rocket. Engineers struggled to design arms that could someday clasp a rocket that was still changing its dimensions, its fuel needs, and its electric power ratings month to month. The company charged with building the arms grew frustrated; they had signed on for X and kept getting requests for Y and Z, and then suddenly back to X.[26]

At the same time, any visitor or engineer moving around the Kennedy Space Center complex had to avoid a sobering, lingering eyesore. The massive crawler, that mechanical beast adapted from coal mining to carry a Saturn V and its nine-armed chaperone, had broken down halfway to the launch pad. The crawler sat there like a sad tractor abandoned by a rural highway.

9

1966—OF SOFTWARE AND STAR BALLS

My father relays a story of a young, obsessive genius who helped mission planning. He lived in a small apartment, on his own. "He didn't have any furniture. Just a few old milk crates and things like that." But this man of simple means, one who frequently forgot to cash his paychecks, was helping correct the computer programs that would take us to the Moon. The young man literally checked the output of Apollo's early computers by reading streams of printed numbers like a broker might have read a stock trading ticker. "In the morning, you'd come in," my father says, "and he'd have left a stack of output, circling each error." Computer experts from NASA and leading consultants at MIT had initially considered the engineer a little too eccentric to take seriously, but they discovered he was always correct.[i]

In 1966, Apollo felt a new bottleneck in a lesser-plumbed realm: computer "software." (The term had only emerged in technical conversations in the late 1950s.) Engineers—primarily a group at MIT's Instrumentation Laboratory—had been so focused on building small, robust, and simple computers for the Apollo missions that they'd neglected the need for programs that would live and run on those computers. At some point, NASA realized astronauts, with little natural interest or ability in using these suspicious new machines, were not going to merrily program them while floating in space. Someone had to create simple, reliable, never-crashing computer programs, with no time to spare.

A freshly minted engineer named Jack Garman joined NASA at this time, having taken a couple of programming classes as college electives. "I had no idea where Houston, Texas, was," he said. "You know, it was kind of 'Alice Through the Looking Glass.' You get in the car, you drive down, and you start going to work. Your eyes are wide . . . everything was very awe-inspiring." He recalled his assignment. "I came in May of '66, and I think back in February of that year the software program for the Apollo onboard computers had gotten in deep trouble." His new employer sent him to five weeks of special classes, to learn Apollo and to learn more about programming. None of his college studies (focused on traditional engineering) had prepared

---

[i] This was John Norton of contractor TRW.

him for the torrent of nonstop novelties, but he was thirsty. "So while I was drinking from the fire hose, I had no trouble swallowing." And, as often happened at a young agency that needed new thinking in Apollo's every nook and cranny, a new employee could suddenly assume huge responsibilities. "So when I got back to the office," he said, "I truly didn't realize it at the time, but I was an expert. . . . [T]hey immediately made me a group leader . . . you know, I'm twenty-one, twenty-two maybe by this time." Soon, just like Marlowe Cassetti, he would be a twenty-something needing to direct people in their thirties.

He encountered reluctance from the older guard. His colleagues said, "What is all this stuff? I just want it to work," Garman recalls. "All this computer stuff, get it out of my hair." The flaky nature of early computers fed the distrust. If the physical switches and wires actually worked (not always certain), then the programs running it—the software—would be full of bugs. And computers were anything but soft and fuzzy. The barrier to interaction was incredible at the time, even once they had *screens* instead of just cryptic strips of paper output. "The displays we looked at were just columns of numbers. . . . [T]hat technology absolutely amazed me, enthralled me, and [just] the ability to stare into a computer," Garman said. "Have you seen the movie *The Matrix*? . . . [T]here's a fellow staring at this sort of waterfall of numbers on a screen." The character is able to recognize, in the cascades of numbers, everyday objects, like a person crossing a street. "That was *deja vu* for me," Garman said. The computer displays of the time only showed a series of numbers (in octal format), and over time Garman recognized enough of the code to nod along, to literally see what the computer was doing. "People would walk up and say, 'You're weird.'"[1]

The problem of computer programming had quietly become central to Apollo's success when NASA made the radical decision to put a digital computer "autopilot" into a flying machine.[2] Garman later explained Apollo's autopilot as follows: "Kind of like if you were walking down the street, you open your eyes to see where you are once every two seconds, and you see the hallway around you and the ceiling and the road ahead, and then you shut your eyes and then decide where to put your feet. . . . So you take three steps. Then you stop and open your eyes, look around, shut your eyes, and go. That's exactly how the navigation and control work, okay? Read all the parameters, do the calculations, and pump out the next two seconds' worth of commands for which way to point engine bells and throttles and so on."[3] The Apollo autopilot used a sensible sort of step-wise flowchart now familiar to much of the world's software.

In May of 1966, a senior NASA leader had just declared that software would soon be the primary pinch point of the entire Apollo program. The *good* news arrived just two months later when MIT delivered a new and improved version of the physical computer itself. Compared to the older model, it ran faster, weighed less, consumed less power, and, crucially, held twice the memory (a whopping two kilobytes). It all fit within the volume of a shoebox. Engineers had also devised a way to make the final version watertight, because an astronaut's space spill could drift within a capsule and ruin exposed circuitry. This computer was, for 1966, absolutely at the bleeding edge of technology, in terms of its reliability, its small, robust packaging, and its inner circuitry design, utilizing chips.

Now the programming, a task not even mentioned in MIT's original Apollo contract with NASA, became job number one. The number of engineers assigned to the software grew and in 1966 finally passed the number working on the computer's hardware. In a mode that would eventually grow to be sacred among programmers, the work took on a round-the-clock schedule. Stacks of program cards piled up for overnight runs, and managers started holding "configuration meetings" on a nightly basis, just to keep track of who had changed what in each segment of the Apollo computer program.[4]

It was an incredibly stressful time for the MIT Instrumentation Lab engineers. NASA managers "descended on us" in Cambridge, Massachusetts, one engineer later recalled. The NASA task masters emphasized that perfect was the enemy of good, that if they were to meet their deadlines, the software would simply not be as good as it could be. It wouldn't beautifully manage fuel supplies and wouldn't achieve a perfect textbook orbit around the Moon—just an orbit that wouldn't kill the astronauts. Management pushed them to turn out version after version of the Apollo software that, as one engineer recalled, "was just so lousy, so full of bugs."[5]

Finally, engineers had to decide where and how to store the computer program on a spacecraft. There weren't many options at the time that were space-worthy. Whatever held the computer program, it had to be completely reliable, 100 percent of the time, no matter how the spacecraft might shake or tumble, no matter if it was hit by a heavy dose of space radiation or extreme temperatures, and so on. At the time, an increasing amount of software and data storage relied on reels of magnetic tape, but these were too delicate for the long Moon missions. Extreme temperatures and a blast of cosmic rays could corrupt or erase the tape, and reading the program would require motors to wind and unwind it. Extra motors, even small ones, meant more things that could leak or break.

The engineers finally selected a type of storage that had first wired the primitive brains of jukeboxes in the 1950s. It was a rock-solid memory that could withstand just about anything, from impact with drunken swing dancers to rocket launches. The Apollo computer program resided, in the end, in a solid mass of wires and little magnetic rings, weighing about three pounds in all. The now-ancient-looking technology relied on a nest of wires, each of which went *through* some magnetic donuts (equivalent to a binary "one"), and *outside* other magnetic donuts (equivalent to a binary "zero"). The term "ropes" emerged in reference to the bundles of wires going in and out of these hoops.[ii]

The memory was difficult to build. Expert seamstresses, dubbed "rope mothers," helped NASA literally sew the memory together into dense mats of thousands of wires and the tiny bead-like magnetic cores. A finished bank of the memory was about the size of an e-reader today, with a tiny cross-hatched weave giving it a metallic plaid appearance. Each Apollo mission would have its own specific computer program and its own specific sewing pattern. Once assembled, it could not be reprogrammed during a mission. At first, some NASA mission planners wanted a *programmable* computer, but over time, many saw the unalterable memory as an asset. Astronauts couldn't mistakenly alter the spaceship's computer code. And once the mothers finished sewing a program, engineers could run tests on it for weeks to make sure it was bug free. Since it was literally hard-wired, as memory, there was no place for new or surprising errors to emerge. Once built, it would only present errors that the mothers or engineers had themselves woven into the wires and donuts.[6]

Half a planet away, Sergei Korolev felt pressure similar to that of America's early software engineers. The Soviet Union had Moon plans of their own. In late 1965, they added to Korolev's already enormous workload by ordering him to design a manned flight that could orbit the Moon. As the calendar turned to 1966, he soldiered on despite failing health. Arriving home after his standard twelve-hour days, he would sit at the entry to his home, just past the posted guards for his apartment. There, talking over the day with his wife, he would slowly build up the strength to climb their stairs. Years of the gulag, and being dragged in and out of Stalin's graces, had aged him.

In mid-January, he attended the Kremlin hospital for what he assumed would be minor surgery (essentially a colonoscopy). The surgeon's work morphed into an eight-hour crisis as he encountered one complication after another in Korolev's compromised state. In the end, the patient lost too much blood, and the Soviet Union lost the man who had engineered so many of their milestones in space. The next day, *Pravda* finally announced the Chief Designer's name and published his photograph for the world to see. A formal state funeral followed in Red Square.[7]

For a time, their space program paced onward with Korolev's momentum. Just a few weeks after his passing, the Soviet Union announced another first: their unmanned probe, Luna 9, had survived landing on the lunar surface. "Landed" may be too gentle a term. Luna 9 crashed into the Moon at high speed, but just before

---

[ii] This gives a simplified version, but for readers who are physics nerds, I will add that the metal rings were circularly magnetized (i.e., clockwise versus counterclockwise). For computer nerds, there was also a small dynamic memory, where the currents could "write" the cores by reversing their clockwise versus counterclockwise magnetic polarities.

impact, it tossed off a lighter, bottom-heavy probe that rolled, righted itself like a children's wobbly toy, and began transmitting to Earth. It sent photographs of the landscape (though unfortunately and unintentionally distorted, making all features 250 percent taller in the vertical direction). Luna 9 appeared to solve the riddle of the lunar surface when it failed to sink into a loose drift of dust, but rather sat on fairly stiff terrain. "The Moon Speaks Russian," Soviet headlines claimed. In the spring, the program placed an unmanned probe into lunar orbit, tightening a long-distance grip on the Moon, in the way that Sputnik had for Earth.[8]

Again coming in second, America put their own robotic lander down in late spring. All on its own, relying in part on my father's calculations, Surveyor I slowed from 6,000 miles per hour to just three-and-a-half miles per hour as it reached the surface. Its three little feet hit nearly simultaneously and then made one light bounce before settling to rest. It began sending a string of ten thousand photographs back to NASA's Jet Propulsion Laboratory in California. Surveyor I was not tiny. About ten feet high and fourteen feet wide, it had solar cells, digging tools, and equipment for chemical analysis, as well as for measuring temperature and radiation on the Moon. Its gear survived the landing in good shape, and its instruments estimated a surface not unlike freshly plowed ground on Earth. All of this was great news for Apollo. Radar-guided landing on skinny metal legs worked, and the Moon, in at least a couple of spots, wasn't covered in quicksand.[9]

Months later, the next Surveyor was on its way, but this one proved again that many things could go wrong on the way to the Moon. One of its engines failed, and it went into an uncontrolled tumble, spinning twice as fast as a standard vinyl record. Engineers never regained control, and the multi-million-dollar satellite, a complete loss, made a new little crater when it reached the Moon.

The prototype for a manned lander lagged behind schedule, despite the best efforts of its contractor parents, Grumman. The lander's solitary ascent engine, charged with starting the journey home, was still acting up. Throughout the year, it kept showing unstable combustion, and NASA asked the subcontracting company charged with this engine, Bell Aerospace, to explore alternatives. But beyond that critical engine and its lift from the Moon, the lander simply suffered being last to the party. Because NASA's decision to have a separate lander had matured fairly late in an already break-neck game, engineers played catch-up to the rest of Apollo. We can glimpse the stress of this, in a world before computer printers, as the Grumman Company struggled to complete the lander's required diagrams and drawings for NASA. Those awaiting the drawings included Houston engineers who had to confirm it could dock properly to Apollo's command module and Huntsville engineers who had to confirm it could fit within a metal shroud at the top of the Saturn V rocket. Lead lander designer Tom Kelly described, at the peak of activity in 1966, having to crank out four hundred large drawings each *week*. They improvised a plywood work surface. "The table was three feet high and about ten feet square," Kelly wrote. "Several engineers or draftsmen worked on their hands and knees on the table, drawing directly on large sheets of vellum paper or on the white painted aluminum sheets."[10]

Engineers in Houston challenged the prototype lander to see if it was steady on its feet. "It was a swing rig, kind of like a pendulum," engineer Tom Moser relays. "We'd pull it back and let it swing and we had a release on it and would drop the model." In this way, using a lander one-sixth of full size, engineers dropped it over and over, at different speeds and at different angles, measuring when it was likely to tumble over, when it bounced dangerously, and when it settled to a nice landing. And they awaited the chance to drop a full-scale lander in a Moon-like vacuum as well.[11]

Now that NASA's big environmental chamber really worked, holding a vacuum without leaks or implosions, engineers began spacecraft tire kicking in earnest. Putting a model Apollo spacecraft in the vacuum taught engineers a great deal. For example, in the command module, engineers watched the urine exhaust lines freeze solid. They started installing extra little heaters to keep the lines unfrozen and altered mission plans to use advantageous turns toward the sun. Windows fogged over to the point of being useless, as the sealants surrounding them outgassed material to the vacuum. In short order, engineers had found more than forty design deficiencies in the main Apollo spacecraft.[12]

Eventually, astronauts inhabited vehicles within the vacuum chamber, risking their lives every bit as much as they might in space. Simulating the conditions of space meant death waited on the other side of a thin metal hull. More than one astronaut later confessed they felt *safer* on the way to the Moon than in the big test chamber. Space was absolutely reliable—one knew what to expect—but a mistake or malfunction in the chamber might create sudden changes to horrible effect.

Engineers at the Cape also ramped up efforts to give astronauts practice and familiarity with the strange missions to come. The game here was to *simulate* the reality of a trip through space as realistically as possible, and NASA started from scratch.

Frank Hughes started an unlikely path into Apollo's inner workings as he suffered through Idaho's coldest days, wondering why he stayed in physics graduate school. Needing respite from the frigid air of a cross-campus walk, he popped into a building and found a NASA recruiting booth folding up shop for the day. He asked them to wait, but the recruiters said he was pretty late at this point. They only had one open slot left, in something called "simulations." Hughes thought about his nearly frostbitten nose, then thought about the weather in Florida or Texas, and signed up on the spot. "I skipped the rest of that night, went home and spent the entire time filling out, with my little portable typewriter, all the information required." NASA sent him to the Cape. Whether or not his new supervisors realized it at the time, Hughes was a perfect fit for the young field of simulations. His physics training and insatiable curiosity primed him to both understand the physical realities of an entire Moon mission (with every jerk, spin, wobble, and shadow) and also help create a convincing practice version for astronauts using a mix of optical tricks and computers.

Hughes found the Cape to be exciting, even too exciting at times. "It's the best thing about working in Florida instead of Houston. At Houston, it's just a white-collar job. Get down to Florida, and it's a dangerous, highly industrial business," he says. "Everything will either burn you, freeze you, poison you, or crush you." He relays a particularly scary event, when a storage tank started leaking liquid oxygen late at night, near the launch pad. A dense fog of super-cool oxygen passed over a nearby roadway. "If you run liquid oxygen across asphalt, it becomes an explosive, but it becomes like a goo." A patrolling security guard drove hesitantly into the fog, having no notion of the leak or danger. "The front end of the car just collapses in this gunk." The man carefully stepped out of the car and took in the other-worldly mess. He slowly backed the car out and, with the running engine just inches above the explosive goo, somehow avoided a more violent ending.[13]

Aside from leaks, engineers occasionally needed to drain excess liquid oxygen. Relatively heavy, the boiled-off oxygen vapor usually stayed in a drainage ditch where it harmlessly dissipated, but on one occasion, a breeze lapped some of it up and over a parking lot. On this dry day, no fog formed. The oxygen invisibly surrounded a number of warm or running cars, and several of them caught fire, sending confused drivers running to safety.[14]

When Hughes showed up for simulations, he found an empty building and a bunch of coworkers, just as new, wondering where to start. To simulate Apollo, they would together need to understand everything about an entire mission to the Moon. They had one complete Apollo manual, such as it was in 1966, and they broke it into several segments to divide and conquer. Hughes, in particular, jumped on the fast track to becoming a generalist. "At the end of the day, everybody put the pieces back together in the book. I'm the only one that said, 'Okay,' and I took it home. I just sat there and read the damn book at the kitchen table." He was unmarried and a long way from home, with no money to spend on a night out, but he loved absorbing Apollo. When he read something he didn't understand, he went right to the experts of any particular system. "God it was great. I not only would call them and ask them how it worked, I'd say, 'Why does it say this?' They'd say, 'Oh, I meant to change that.'" No detail was too small once Hughes drilled into an Apollo system, and the simulation team quickly built an uncommonly broad expertise.[15]

The eventual Apollo simulators—in time, NASA built multiple simulators for both the command module and the lunar module—included realistic cabin interiors, but these were surrounded on the outside by ungainly boxes and cables. The simulated experience of space travel married computers tracking the "ship" in space, whether it was circling Earth, circling the Moon, or somewhere in between, with a series of clever visual effects.

When an astronaut peered from the small windows of a simulator, he viewed a believable image carefully assembled using models and old-school optics. The engineers made a credible model of the Moon (to the extent that its terrain was understood at the time). This large, curved Moonscape lived upside down, with cameras panning over it from below, so that it wouldn't accumulate earthly dust.

The simulators required a realistic star field, dialed to the correct portion of the heavens at any instant. To project an accurate spray of stars on a dark background, engineers used a "star ball": a pitch-black sphere about two-and-a-half feet across, covered with 999 tiny steel balls, machined to be an appropriate size for each star's respective brightness. The complete star ball then became a small globe of the heavens, creating a star field by reflection. When hit by bright light, it reflected a patch of the heavens onto a screen. The tiny steel ball bearings were even colored slightly to match the hue of each star, be it more orange or blue (see Figure 9.1).

If the astronauts rotated their simulated spacecraft, the simulator itself stood still, but the star ball would spin to give the astronauts a correct view of the cosmos. "They

FIGURE 9.1 A star ball used to paint the heavens for Apollo simulators. (Photograph courtesy Frank Hughes.)

were just incredible, gorgeous stars," Hughes said. "So much better than anything we've done [since]." He notes that in a digital simulation, tiny stars eventually have to be either one pixel or two pixels, with no in-between.

When astronauts wanted to practice docking in orbit, a realistic small model of the target ship would give them a lifelike view from their windows, set against the stars. The models made for a few moments of play as well. The television series *Star Trek* premiered in 1966, with its optimistic (and increasingly quaint) view of a unified Earth sending explorers into the cosmos. The creator and cast once visited the Cape and sat in the simulator with an astronaut. "Okay, now we'll turn around and dock," their host announced. As their view rotated, they were delighted to see, instead of a lunar lander, a model of the Starship Enterprise, lovingly assembled by one of the engineers. Using the simulator, the crew "flew" the Apollo command module carefully around the starship. (Hughes claims this influenced the first studio Star Trek movie, where the captain, for no apparent reason, takes his crew on a little shuttle craft, up one side of the Enterprise and down the other.) During this visit, however, the simulated sunlight overwhelmed the dime-store plastic starship. "I looked up," said Hughes. "I'm just on the simulator console, running the thing . . . and the Enterprise was melting. The front end was just kind of drooping down."

Hughes and his young colleagues also pranked astronauts on occasion, using an innovation premiering in the 1960s: invisible tape. They snagged a Florida cockroach, and "we'd tape it to the lunar surface . . . about two inches away from where they're going to land. It's still alive, and it's just trying to get away . . . antennas waving and the legs are darting." Astronauts learned to shake their heads at the occasional goofiness, but their practice schedules were getting increasingly serious. NASA wanted to show the astronauts every conceivable eventuality, to give every possible problem or emergency a trial run before putting their lives at risk. Eventually, the availability of the precious simulators dominated the schedules of astronaut training.

Some parts of astronaut training took them away from the cramped simulators and even away from NASA centers. Hughes recalls taking the astronauts, with no equipment, to the middle of nowhere in Florida. After some time in a planetarium, he tested the men out of doors. "Can you still find Canopus, and Altair, and Polaris, and every one of the fifty-one stars that you had to know?" This was no sightseeing tour. If an Apollo mission lost contact with Earth, the crew would need to navigate home using their reference stars like old-time mariners in wooden ships.

Even if they learned their stars, challenges would await them. A ship orbiting Earth is always on either the bright side in sunlight or the night side in Earth's shadow. On the bright side, the sun lighting up the atmosphere obscures most of the stars. But, Hughes said, "if you get into the shadow of the Earth, you look up, and you cannot see the constellations, because there's so damn many stars in-between all those bright ones. . . . You see more than ten thousand stars."[16] For a stargazer on Earth, the atmosphere dims the star field, hiding thousands, but once in space, assuming a clean window, a viewer takes in the unfiltered version, so complex that it looks alien. And astronauts on their way to the Moon eventually found new complications. Ice crystals (or urine crystals, in some cases) sparkled outside the Apollo spacecraft, and with no rushing air to pluck them away, the crystals would move along, maintaining orbit with the spacecraft and sparkling with reflected sunlight like so many unwelcome stars.

The year 1966 witnessed some tragic fallen stars for manned spaceflight. In February, two astronauts prepared for a Gemini mission. Hoping to check on their capsule assembly, they took a two-seater supersonic jet from Houston to St. Louis. Arriving in snowy, foggy weather, they tried to land using their instruments but crashed and died on impact, as nearby McDonnell Aircraft engineers worked to complete their Gemini spacecraft.[iii]

NASA nearly lost two more in a subsequent Gemini mission. The astronauts successfully docked their Gemini capsule with an unmanned target vehicle in Earth orbit.[iv] But before they could celebrate, the conjoined craft suddenly began spinning. Assuming the target vehicle was responsible, they separated from it, but their own capsule's unwanted spinning only increased. The capsule's interior began to resemble a dryer drum, tumbling one full turn every second. The astronauts' heads banged together. Concern in Houston's Mission Control Center steadily climbed, as the astronauts flew about Earth, moving out of communication range with one Earth station and then, many long minutes later, into range with the next station. At each step, astronauts reported faster spinning, and nothing they tried would control it. Their vision started to blur. NASA knew that, if the two men blacked out, they would probably die in orbit. And since the capsule's radio antenna was also gyrating like crazy, communications became almost impossible. At the last instant, astronaut Neil Armstrong decided to turn off the entire primary system of thrusters. Using a secondary set normally reserved for their descent from orbit, the crew then carefully snuffed out the crazy rotations.

This mission easily marked NASA's worst emergency in space yet. Engineers subsequently diagnosed the cause: after suffering an electrical short-circuit, a thruster found itself permanently in the "on" position. Without a symmetric thruster fighting it, the broken one madly increased the spin of the ship.

[iii] Astronauts Elliot See and Charles Bassett died February 28, 1966.
[iv] Astronauts Neil Armstrong and David Scott survived the troubled flight of Gemini VIII on March 16, 1966.

"I was surprised at that one," Pohl said. "Back in that day and time, our transistors weren't very good." He still recalls the diagnosis, stepping through how a crack allowed some of the caustic, hypergolic fuel to leak into the electronics, causing the short-circuit. Engineers working in Henry Pohl's group redesigned the electrical connections so that any future short-circuit would leave a thruster set to "off."[17] But those thrusters had another problem lurking, as yet unfound, and this one could destroy spacecraft in an instant.

By 1966, Pohl managed an increasing number of people and projects. He recalls when Max Faget's right-hand man approached him about increasing his responsibilities. "I said, . . . 'I don't know how to manage people,' and he said, 'Yes you do. I have been watching you.'" Now Pohl oversaw several systems on each module for the Apollo program and he'd also taken on problem areas of the more immediate Gemini program, like the thrusters.

The hours were incredibly long, and Henry never got to spend the time on his parents' south Texas ranch that he'd planned. He recalls evening work dinners with engineers from various companies charged with perfecting different hunks of the spacecraft. "I could learn more over the dinner meal than I could all day in those formal meetings," he said. The engineers left decorum at their offices and started solving problems on napkins. "We always had Happy Hour before dinner." Most of the assembled opted for cocktails or beer, but not Pohl. "I would order a big glass of half and half. That was half coffee cream and half milk, because I wanted to keep my brain clear, and we would discuss the problems."[18]

Pohl, over several interviews, returned to a 1966 test again and again. "The one that really haunted me," he said. "I still shudder when I think about it. That almost got by." At issue were some titanium tanks that, much like a gas tank in a car, stored an explosive fuel for the Apollo control thrusters (and those for the Gemini capsule as well). These tanks were modest cylinders, about half the size of your arm. Pohl's team ran a long duration test for three of the tanks filled with pressurized fuel and placed in a vacuum chamber. "Twenty-eight days in to that thirty-day test, one of those tanks developed just a little leak," he said. "It was just bubbling out a little bit through a crack." It didn't seem like a disaster. The tank manufacturer quickly diagnosed it; one of their workers hadn't been using gloves, and they believed the natural oil of his fingerprint led to a chemical reaction inside the titanium tank. Pohl was satisfied, but two of his youngest engineers kept coming to see him. They told him the company had to be wrong—there was no way a fingerprint caused it.

"Why not?"

"It would take a monkey to get his hand in there!" they said. The tanks were too small to have a human hand working inside them.

Pohl initially told them to calm down. They had a million things to worry about and the tank explanation looked good enough; the company building it had the right expertise. But the junior engineers wouldn't let it go, and finally Pohl asked the tank maker for further tests. The company collectively rolled their eyes, but they wanted to keep Pohl and NASA happy. This time, they ran ten of their fuel tanks, under vacuum, for a long test. And their workers used gloves throughout the construction of those tanks.

"I think about seventy-two hours into the test, one of those tanks . . . just busted wide open," Pohl said. They had their answer now—there was a major problem. They called off the test, but "before we could get the pressure down . . . two more of them blew up. That was three out of the ten." In an actual Apollo mission, seventy-two hours would have had the spacecraft sidling up to the Moon. If Pohl had not ordered the extra test, one of those tanks probably would have killed three astronauts. That he almost didn't listen to his junior engineers still troubles him.

The actual cause of the problem was an obscure and cruel trick of chemistry. The fuel producer had innovated a new processing step to make the fuel incredibly uniform and, ideally, more reliable—each ounce would be like the one before. But the process left excess hydrogen dissolved in the mix, which in turn degraded the titanium tanks.

This very technical example is illustrative of two things one encounters over and over in talking with Apollo engineers. First, they still feel deep responsibility for their parts of the missions and suffer the mistakes or near misses fifty years later. Next, the culture, by all accounts, encouraged anyone of any rank to speak up if they saw a problem or had a different take for improving the project. As one engineer put it, they were all equals at some basic intellectual level: "If you could communicate it, then that day you wore a suit and you presented it."[19]

While no fuel tanks blew apart in space, America's second spacewalk nearly claimed a life that summer. Unlike its first spacewalk, which amounted to blissful floating, NASA now asked an astronaut to work: to exit the capsule, climb to the back of the craft, adjust a few instruments, and then clamber slowly back to safety. Simple. But (as the Soviets had already learned and kept to themselves) the pressurized suit became a rigid balloon once in space. The astronaut later compared it to hardened plaster of Paris. "It did not want to bend at all, anywhere." Every movement took Herculean effort. His heart rate leapt, he soaked his inner clothing with sweat, and his quickening breath fogged his facemask. He floated as if mummified, blind and clawing at the Gemini capsule's hull. When he yanked on his umbilical tether, it started a horrible physics experiment. Instead of just pulling him toward the ship, it caused him to start spinning, and the tug started the spacecraft slowly tumbling as well.

The astronaut, with his last energies, somehow made it back into the capsule with the help of his fellow astronaut.[v] He later compared it to "putting a champagne cork back in the bottle." He'd lost over twelve pounds in two hours of exhausting space activity, mainly trying to bend the fingers, elbows, and knees of his suit. His visor was completely fogged over, and pints of sweat sloshed around in his suit. He told his fellow astronaut that his back felt like it was on fire. The Gemini spacesuit (not made by the Latex Corporation) had started to tear in back, and the unshielded sun had branded him with a second-degree burn.[20]

None of this boded well for the notion of an astronaut descending a ladder and walking on the Moon. Engineers began designing a thin layer of circulating water that could sooth extreme temperatures in the Apollo suits.

Unfortunately, everyone understood that worrying about a Moon suit was a luxury of sorts. As with so many Apollo projects, it required one four-hundred-foot-tall leap of faith. None of NASA's preparations meant a thing if they didn't have a rocket that could propel Apollo to the Moon.

In 1966, the Huntsville team hustled toward a maiden, unmanned launch of the full Saturn V rocket, slated for early 1967. A new test facility for von Braun's empire had sprouted along the Mississippi River. After relocating some 850 families from the rural swamplands, and with no major cities nearby, the engineers hoped loud rocket tests wouldn't cause much trouble, and a barge could easily haul hunks of a Saturn rocket downriver, to the Gulf of Mexico and on to Florida. A chief administrator of the new Mississippi test facility said the pace of work was like "riding with one foot on each of two galloping horses."

In April, the new facility had enjoyed one of the few successful tests of the troublesome second stage of the Saturn V, where a hydrogen engine ran for a clean fifteen seconds. But just a month later, a sloppy shift change led to disaster. One crew of technicians turned off pressure sensors to the fuel tank, but they failed to inform the next shift. The incoming technicians put helium gas into the fuel tank to check for leaks, and, with no warnings from dormant sensors, the pressure kept building until the enormous tank exploded, injuring five employees. In the tank's post-mortem, engineers found tiny cracks that predated the explosion. Checking other, similar tanks, they found these filigree flaws to be widespread. The calendar showed only so many days to keep making these trips to the drawing board.[21]

Engineers continued scrutinizing the Saturn V's weight like a ballet choreographer narrowing eyes at his principal dancers. Every pound engineers could remove from the rocket helped them get a few extra ounces headed for the Moon. In the

[v] Astronaut Eugene Cernan toiled outside the Gemini capsule for about two hours on June 5, 1966, and was helped back aboard by Mission Commander Thomas Stafford.

never-ending pursuit of weight loss, engineers now turned to the already skinny second stage. Trimming gristle here was more efficient than trimming the first stage. Since the second stage spent more time on the mission than the first (even just a few extra minutes during launch), the accounting of multi-stage rockets meant every pound from the second stage was worth more than a pound trimmed from the first.[22]

Making the stage's walls as thin as possible exacerbated the ongoing troubles with the very long, precisely welded seams. Sensing urgency, Wernher von Braun pulled together a select engineering group to figure out a more reliable welding process. With super-cold hydrogen fuel and persnickety metals in ever-thinner sheets, the problem looked nearly impossible. After deliberating, the engineers recommended changes to the welding workshop's environment. Decreasing the humidity in the manufacturing facility would help, and workers also needed to remove any chance of airborne impurities, like dust motes, lighting on the metals before welding.

The resulting solution was a so-called "clean room" environment, made more famous in modern times in microchip processing and nanotechnology, with workers in surgeon-like suits, masks, and booties. Engineers reworked the manufacture of the second stage, conditioning the air to cut humidity in half, and having workers submit to ultra-clean standards as they moved into the facility. They passed through an airlock to minimize whatever unclean puff of air might enter with them through a traditional door. They wore gloves and smocks designed to minimize lint. They ran their shoes through a special scrubber on their way into the airlock. Once inside, they found squeaky-clean epoxy floors, mopped obsessively. The bright white walls and their garments made them look like extras for George Lucas's early science fiction dystopia, *THX 1138*. Though problems persisted, these careful steps began to pay off by the end of 1966, with stronger, more reliable seams on the second stage. There was little time to spare if they wanted to launch in early 1967.[23]

Pieces made their ponderous way to the Cape. Like a bizarre religious rite, a barge floated the first stage, under an enormous shroud, down the Mississippi, through the Gulf of Mexico, and around the tip of Florida. The second stage shipped from California and through the Panama Canal. The third stage made its way from southern California by air. To maneuver it to the airport, engineers commandeered entire roads and employed an eleven-ton transport vehicle. In one case, they had to wait for local utility companies to bury some power lines that otherwise would have snagged the rocket segment. And despite moving just a few miles per hour, the transporter ran over an even-slower skunk, perfuming the truck but leaving the rocket stage unscented. Despite being the smallest of the three, the third stage still required extreme measures for air transport; enter the "Super Guppy," a cavernous plane that could swing open its head, like some sort of outsized Pez dispenser, to allow rocket stages on board (see Figure 9.2).[24]

FIGURE 9.2 A "Super Guppy" open for loading in 1965. This custom airplane transported the third stage of Saturn V rockets and various Apollo modules to Florida. Once aloft, it resembled a flying sperm whale. (NASA photograph.)

Wernher von Braun's time in Huntsville had been meteoric. The city had grown from about 16,000 souls in 1950 to nearly 150,000 in 1966. He wanted his program and the surging Rocket City to continue. The heavens were so vast that they required armadas of spaceships and the rockets to lift them. Yet, even as he watched his ultimate rocket slowly come together, he was more worried about the future than ever. His visions would require seas of money, and by 1966 he sensed the momentum slipping. After years of work and advocacy to build a lasting space program, he watched tepid politicians and worried voters start to avert their eyes. In their nervous shifting, he recognized an existential threat. America might stop at the Moon, if it even got that far.

That year, he became especially blunt when speaking in public. "Our main effort today is busily *destroying* the very capability that we have built up to put a man on the Moon," he said. He wrote an article for the *Los Angeles Times* that summer, telling readers "to make a one night stand on the Moon, and go there no more, would be as senseless as building a locomotive and a transcontinental railroad, and then making one trip from New York to Los Angeles."[25]

Lyndon Johnson had started squirming over funding. The war in Vietnam showed a growing monetary appetite, and his domestic programs put a new strain on the federal budget as well. Many of his advisors told him cuts must be made and taxes would have to creep up at some point. By the end of 1966, he agreed to cut about $1.5 billion from his next budget. Meanwhile, in Congress, more and more members looked skeptically at NASA's funding. Congressional representative Donald Rumsfeld of Illinois tried to cut Faget's lunar receiving laboratory, since it looked like an attempt to perpetuate space exploration. Its $9 million survived deliberations, but soon Congress reduced a post-Apollo program called "Apollo Applications." The nation's vision turned increasingly earthward.[26]

Humanity's troubles could now visit any living room via live satellite broadcast or quickly prepared clips from around the world. At home, where cities once wondered if a "systems engineering" approach could cure urban ills, decaying city centers fumed. Confidence in the nation's path, in the federal government, and in space-age technology began to ebb.

Presumptions of a united ramp of national progress splintered. The Student Nonviolent Coordinating Committee, a cornerstone of the Civil Rights movement, with its logo of white and black hands joined in solidarity, voted in late 1966 to bar white members. Martin Luther King Jr. now sometimes found his nonviolent speeches greeted with public boos. Voices fed up with the lack of progress toward racial equality shouted "black power!"[27] Watching a confusing war abroad and racial tensions rising at home, some youth began wondering about more than government. Maybe the way society pieced things together was outdated or even rotten from the roots. In the rising "Crisis of Authority," youth considered that even their parents and churches might be wrong about things. *Time* magazine published its infamous "Is God Dead?" cover in the spring of 1966.[28]

The engineers of Apollo granted themselves little time to worry about these shifts and movements. In a rocket's ivory tower, tanks could leak, computers could fail, hulls could crumple, and engines could shake to pieces. When the engineers awoke in the wee hours, they stewed over these technical problems rather than societal ones.

Having four hundred thousand human beings join a multi-faceted, ambitious project, with dozens of question marks blocking the finish line, created an understandable babble of voices, from all levels, all facilities, and from scores of contracting companies. Many voices expressed worries, but conflicting opinions were plentiful for most any subtopic. Well-informed warnings, along with intelligent rebuttals, were commonplace. And some nagging problems faced dire predictions no matter which way NASA chose to solve them.

Floating through this noisy storm, a couple of late-1966 voices worried about the interior of the main Apollo spacecraft. Henry Pohl visited the craft's California production plant that year. He "crawled in that Command Module just to look in there," but was struck by all the doo-dads that were added to support the astronauts. "I had absolutely no responsibility in that area at all, but I says . . . 'You can't put all of that stuff in there. It'll burn.' . . . [T]hey had it full of what I thought was combustible stuff." He did make a phone call the next day, but let it drop. "There was another one of those instances where a light bulb lit up, and I didn't follow up on it."[29]

But Pohl wasn't alone. A NASA administrator named Joe Shea received a September letter from an executive at General Electric. The letter set out serious concerns for the 100-percent oxygen atmosphere. It acknowledged that using pure oxygen in the capsules for Phase I (Mercury) and Phase II (Gemini) had worked fine so far, but they were just one spark short of a disaster. "The first fire in space," read the letter, "may well be fatal." Shea asked one of NASA's test divisions to review the amount of flammable material within the capsule, but they didn't reply for seven weeks. The issue was just one of thousands of concerns. Here, engineers were trying to put three men on top of the most powerful, complex, and explosive rocket ever conceived. And some administrator was worried about Velcro straps in oxygen. But they did eventually review the cabin and reply to Shea, saying hazards of fire within the Apollo craft were "low." Shea forwarded their report to the executive at GE, and he hand-penned a little post-script. "The problem is sticky—we think we have enough margin to keep fire from starting—if one ever does, we do have problems."[30]

By the end of 1966, the world had a true glimpse of wonder, with NASA's second remote probe entering lunar orbit. The robotic craft dropped to just thirty miles above

FIGURE 9.3 The Moon's Copernicus Crater as viewed by Lunar Orbiter II. This 2013 version greatly improves the version released in 1966, thanks to the Lunar Orbiter Image Recovery Project. Like the Soviet Luna probes, the Lunar Orbiters had to develop their own film and scan the images before transmitting the results to Earth. (NASA/LOIRP photograph.)

the lunar surface, as it tried to understand the precise nature of the Moon's gravity. NASA probes had found evidence by now that the Moon was lumpy under its surface, with some very heavy areas, or "mass concentrations," creating regions of stronger gravitational pull. These clumps complicated any Moon mission, since an orbiting spacecraft would find the Moon tugging harder in some stretches of the orbit than others.

The low approach made for a wonderful photo opportunity. Before it eventually crashed into the Moon, the probe radioed back bushels of gravitational data and a striking landscape photograph. Some in the press called this the picture of the year, and others upgraded it to the picture of the century. It was one of the clearest, closest lunar photographs yet. The mottled and pock-marked surface of the Moon became much more tangible. Instead of flat circles and patches through a telescope lens, people could now see harsh ridges, valleys, hills, and long shadows—a dramatic land-scape but not what anyone would call welcoming (see Figure 9.3).[31]

10

1967—FROM MADNESS TO MIRACLE

"Hindsight is wonderful," Max Faget said of an early 1967 Apollo test. He noted that they'd never had a problem with oxygen inside their capsules—not even a worrisome hint of trouble in the first five years of space capsules. "But, you see, after I started thinking about it, kicking myself for being so stupid," he said, "I realized that the difference between Mercury and Apollo was that one Apollo was probably equivalent to maybe twenty or thirty Mercuries, simply because there's so much more volume in [the Apollo capsule] and there's so much more stuff."

"It was a schoolboy mistake," said Faget's lifelong design partner Caldwell Johnson. "And, you know, a bunch of us were all involved in it." He pointed out that the problem could only have happened for a ground-bound test. Once away from Earth, the Apollo missions would fill the crew capsule with oxygen but only to one-third of a normal atmospheric pressure. For test runs on the launch pad, the engineers kept the astronauts stewing in an entire atmospheric pressure's worth of pure oxygen. As Johnson put it, "It was a bomb ready to go off."[1]

Many space age accounts claim that the space race had functionally ended by 1967: that after losing Korolev and watching America make progress, the Soviets had just shrugged and abandoned any Moon ambitions. The USSR eventually claimed as much, and by the close of Apollo, this idea grew to be gospel. (As newsman Walter Cronkite said in a 1974 television special, "It turned out that the Russians were never in the race at all.") But the truth is not so straightforward. Records show that cosmonauts practiced special helicopter landings throughout the 1960s (presumably to simulate Moon landings), and the Soviets continued testing Korolev's massive N-1 Moon rocket through 1972. In any case, if America was not in a race, that was not at all clear to the engineers in 1967. A Soviet Zond spacecraft had already flown, unmanned, past the Moon and on toward Mars. This new, larger ship could presumably hold a few cosmonauts at some point. The fiftieth anniversary of the Russian Revolution loomed in October 1967, and NASA assumed the USSR would try for a new milestone, perhaps taking a crew of cosmonauts for a Zond spin around the Moon.[2]

NASA wanted to launch an Apollo mission by the end of February, despite numerous troubles. In preparation, they needed to test two modules of the eventual

Apollo spacecraft—the command module, a larger, more sophisticated version of the earlier capsules, where the astronauts would ride to the Moon, and the service module, nestled right behind it. One could simplify describing the service module as just a large can filled with fuel and oxygen—a sort of U-Haul trailer for the Moon missions—but it handled a great deal more than that. Apollo's service module featured the primary engine for the lunar flights, and Apollo used that engine both to achieve a parking orbit around the Moon and then to jumpstart its trip back to Earth. Not built for glory, a Moon-bound service module would never bask in a parade or museum. Before re-entering Earth's atmosphere, the command module jettisoned its companion on each trip (see Figure 10.1).

In January, since the Saturn V was still coming together, cranes loaded the two modules carefully on top of a lesser Saturn rocket. This earlier model would serve as a kind of stunt double for the Saturn V during the launch pad tests.

The engineers worried and fussed. Two leading managers had a "knock-down, drag-out" January argument over the crew compartment, which still seemed a bit haphazard. It had accumulated more and more material inside; straps, netting, and

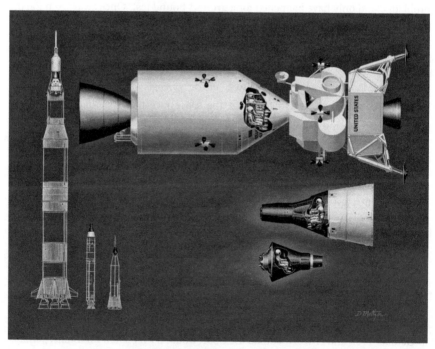

FIGURE 10.1 Artist's rendering of the Apollo modules (*top*) and the Saturn V rocket (*left*), compared, respectively, to spacecraft and rockets from earlier phases of the manned space program, Gemini (the two-seater) and Mercury (the one seater). Running left to right across the top are the service, command, and lunar modules attached in the way they would eventually fly to the Moon together. (NASA image.)

tangles of wire crossed and circled one another, hanging from the sloped capsule walls and even snaking under the astronaut's seats. Preparing for a January test, one astronaut requested that engineers leave behind a polyurethane pad; he said he needed it when he stood up to open the hatch from within, because otherwise he would be standing on a bunch of loose wires.[3]

On January 27, NASA planned a simulated countdown. The rocket would sit empty of fuel, but the astronauts would ascend an elevator in the support tower, walk across a gantry far above the Florida landscape, and climb one by one into the capsule. There, in a cabin atmosphere of pure oxygen, they would test a few critical systems and practice the countdown.

One of the busybodies monitoring the test was Joe Shea, who had handled the warning letter in 1966. By January of 1967, Shea had emerged as something of a NASA celebrity, nearly rivaling von Braun. He was handsome, well-spoken, and quick-witted. Known for bright red socks and plentiful puns, he became a go-to person for the media. *Time* magazine had him queued up for its cover, once NASA finally launched an Apollo mission.

On the morning of January 27, Shea communicated back and forth with senior astronaut Gus Grissom, who sat inside the capsule. In cabin tests a week earlier, astronauts had complained of a sour milk smell, perhaps from something burning, and when they'd plugged their suits into an electrical circuit, something in the air had made their eyes sting. Today, the audio lines between Mission Control and the capsule were terrible. Voices cut in and out, fighting static. "How do you expect to communicate with us in orbit if you can't even talk to us on the pad?" Grissom asked Shea. He suggested that Shea come up to the capsule himself to see if he could make sense of which wire was snagged on what to foul the signal. "It's really messy," the astronaut said. "We want you to fix it." Shea considered it, and even inquired with other staff to see if it could fit in there with the three astronauts. There was room for him to sort of lie at the astronauts' feet during the test, and maybe he could find the problem. Other engineers said it was too much trouble to get him to the pad and up the tower; meetings in Houston expected him soon anyway.[4]

The Cape was learning what eventually came to be known as "Saturn V minutes," where complexities of the rocket and the Apollo craft meant any test or countdown took about five times as long as anyone would have predicted. The astronauts waited and fiddled with controls, communicating with the engineers as best they could.

Then, across the static-laced line, engineers heard a concerned voice. "There's a fire in here."

"We never did find out what caused it," Faget said later. "It only takes a teeny bit of stuff, with some teeny bit of flammable material. . . ."

Frank Hughes, not even a year into his NASA work at the time, recalls being in the trenches that day. He was working a 3:00 p.m.-to-midnight shift to revamp one of the simulators. As dusk arrived, a colleague ran up to him. "Something wrong with the crew? Have you heard anything?" He hadn't.

"So I try to walk in and call, and the phone was ringing. From then on, it was all hell broke loose," Hughes said. "The crew was injured; at first, that was what we knew, that they had an accident. Then it just got worse, and worse, and worse."

The lives of three astronauts ended within a terrifying half minute.[i] Something within the cabin sparked—most probably a fatigued, unprotected wire—and with pure oxygen inside, flammable material lit up like dry grass. The astronauts called for help. They tried to pull the hatch inward, relying on months of training. But the fire stoked the cabin pressure higher, holding the hatch firmly in place. With a sickening explosion, the fire finally ripped through the side of the capsule, knocking back technicians who had scurried to help. Contrary to subsequent grisly media reports, the astronauts were not burned to death; all three died from inhaling the fire's carbon monoxide.[5]

The nation mourned for its astronauts, but it also turned a skeptical eye on the space program. On the heels of NASA's eventual public accident report, the *New York Times* ripped the agency, saying, "even a high school chemistry student" would avoid a 100-percent oxygen atmosphere. It leveled charges of "complacency . . . incompetence and negligence."[6]

Had the program just been in too great a rush to launch within early 1967? "No, no. I didn't think so, no," Faget said. "If they'd said, 'Relax, take another three or four months,' we'd still have probably flown the same spacecraft, still would probably have run the same tests, still probably had the same goddamned fire."

In fact, the Apollo test in oxygen wasn't the only one of its kind, and the accident didn't quite put a stop to the practice in time to save other lives. Engineer Larry Bell had been working on spacesuits for the Gemini program. He recalled a "very, very bad day in my life" shortly after the Apollo fire. He received a phone call from the Brooks Air Force Base in Texas with bad news about a suit test. In a much quieter tragedy, two airmen died when fire burst forth in a pure-oxygen chamber. On his way to the base, Bell remembered that their temperature-control unit in the suit had magnesium in it. "I thought, 'Oh, my word. Magnesium burns like a torch.'" But in the end, the suit wasn't the culprit. Exposed wires in the chamber had sparked the fire, just like they had at the Cape.[7]

---

[i] Gus Grissom, Ed White, and Roger Chaffee. The tragic test became memorialized as the first official program mission, Apollo 1.

The Soviets, who sent their sincere condolences after the launch pad tragedy, suffered their first cosmonaut death soon after, and it marked our species' first death during a space mission. A cosmonaut took the new Soyuz capsule into orbit in April, where he encountered nonstop problems, from maneuvering to communications with the ground. Attempting to come back to Earth early, his capsule began tumbling and the Soyuz rammed into Earth at deadly speed.[ii] (The Soviets had never embraced the softer, water-based landings, in large part to maintain secrecy.) After losing four spacefarers in a matter of months, the world owned a deeper appreciation for the dangers of leaving our natural home.

The Apollo tragedy sank spirits throughout NASA and in the ranks of its devotees. As authors Murray and Cox penned, "It was not only three astronauts who died. . . . Some of the program's lightheartedness and exuberance died too."[8] Prospects for fulfilling Kennedy's promise looked grim. While Max Faget, Henry Pohl, and countless others flogged themselves with what-ifs, Joe Shea—he who had fielded the worried 1966 letter and had *almost* been in the cabin with the three astronauts—accepted blame on a personal level. His star faded as quickly as it had ascended. His work ethic, always beyond reproach (if not downright frightening), became untenable. He worked around the clock reviewing accident diagnostics and plans for changes. As many engineers did that spring, Shea turned to alcohol to end one night's work and stimulants to face the following early morning.[iii]

Shea began to worry NASA's leadership. They decided he wasn't fit for congressional testimony and they urged him to take a leave. He refused. As the investigation plodded from one agonizing week to the next, he grew "increasingly melancholy" according to one observer. Then, at a Houston meeting, Shea rose and started giving an update on the investigation. He started calmly, "but within a minute, he was rambling. . . . Whatever was happening in Joe's head, it all came out in a jumble of mixed words and meaningless sentences," engineer Chris Kraft later wrote. By July, Shea resigned from NASA.[9]

The full investigation, led by Faget, Shea, and others, never pinpointed the exact start of the blaze but clearly implicated the problems of flammable materials resting in pure oxygen gas. With machine-like efficiency and determination, the engineers made a series of recommendations for the Apollo spacecraft. Some were obvious, like reducing flammable materials and including nitrogen in the cabin atmosphere for all ground-based tests, but others fixed long-simmering problems, unrelated to the fire.

[ii] Vladimir Komarov died on impact on April 24, 1967.
[iii] During these months, one young engineer named Larry Moran died from, as best as anyone could tell, overwork and poor nutrition. Coworker and chief lander designer Tom Kelly blamed it on the man's workload and obsessive dedication.

"We put 125 things on that list for the command module," engineer and director of flight operations Christopher Kraft later relayed. "Within six months, before the end of 1967, we'd done all 125. Then we did the same thing for the lunar module."[10]

Optimism steadily returned to the Houston center as the program gathered itself and dove more obsessively than ever into the work. A number of the engineers say that the disaster, a clear demonstration of the stakes, drew a new level of focus from younger employees in particular. As von Braun said in one of his pep talks, "I think we should all understand that we are not in the business of making shoes." The nature of the work, its aspirations, and its risks were wholly unique.

Bits of good news emerged. Programmers had solved most of the software dilemmas for the Apollo computer by the spring of 1967. And the unmanned probes were perfecting trips to the Moon, whether orbiting there to better map the terrain and gravity, or actually landing. In April, Surveyor 3 inspired the space program when it made a bouncy but successful landing, then used a robotic arm to probe and scoop the lunar soil. To further break the depressive clouds, Houston's NASA leadership decided to hold a center-wide party in May, celebrating the sixth anniversary of lifting the first American to space and back.[11]

In retrospect, many engineers spoke and still speak of the tragedy as not just a turning point but also an important part of the Apollo process. "You kind of hate to say it," Henry Pohl allowed, with a pause, "but it did give us the breather that we needed to fix an awful lot of problems in the program that, as bad as that was, may have been worse than that, because if we started on a journey to the Moon," a cabin fire in space would have been even more devastating to the program, with no remnant evidence to evaluate.[12]

NASA grew ever more confident in their revamped command module as a ride to the Moon, but it wasn't clear, in 1967, if astronauts would have anything that could then get them to the lunar surface. The lander was still an earthly mess.

Pohl recalls a Houston-based test, putting the lander into the big environmental chamber, lowering the pressure, and then mimicking the harsh temperature dichotomies of space. "When we . . . started shining the simulated sun on one side and the other side exposed to liquid helium temperatures," he said, "it was literally tearing itself apart." The warm-side metals expanded, while the cold-side metals contracted. Engineers eventually solved this by coating the structures with reflective metallic foils. The lander had never been pretty or sleek, but it came to resemble a child's awkward, homemade Christmas ornament.[13]

In late June, Cape Canaveral received a completed lander for inspection and testing. It was four months late, but engineers still hoped to get an entire Apollo spacecraft in orbit by year's end. NASA's director of Apollo launches wrote a letter to the contractor, Grumman, in charge of the lunar module. "That LM you sent us yesterday is supposed to fly in space, but I wouldn't even allow it on the launch pad," he wrote. "It's propulsion tanks and plumbing leaked like a sieve . . . there were sirens wailing everywhere." He called the delivered craft "a piece of junk, garbage." Grumman opted not to debate him, slinking away with their rickety project in tow.

A new phenomenon emerged for the poor lander. The drive to make it as light as possible—that was the whole advantage for having a separate lander, after all—meant that many of its parts were carefully machined to be as thin as possible. In some cases, engineers used a special chemical process to thin the walls and legs, but, unbeknown to materials scientists at the time, the surviving surface endured a new kind of metal stress. By mid-1967, engineers started finding a slew of cracked parts all over the lander, be it on the inside, on the thin hull, or in one of the legs. Lander architects pivoted to use a newer form of aluminum, and they instituted pervasive inspections, going part by part instead of lander by lander.

Wiring posed another nasty problem. "Probably the worst choice I made," according to chief lander designer Tom Kelly, was the use of very fine wire, with tiny connectors in all circuits. The lander "had many miles of such wiring, so this one item saved hundreds of pounds—but at the cost of recurring wire breakage" and difficulties with electrical connections. These troubles persisted through 1967 and beyond. Never mind the paranoid vision of a tiny space pebble punching a hole in the paper-thin module—the nightmare of it short-circuiting and stranding astronauts on the Moon looked completely realistic.[14]

My father also worried about the lander but less like a doctor and more like a parent prepping a child to leave the house, making sure to pack the lander's lunch box. He worked non-stop on mission planning after a recent promotion. "I headed the 'consumables' effort," he says. "We had to calculate the use of oxygen, water, carbon dioxide counters, propellant, and so on. . . . [W]e flew simulated missions [using computers] and used these to calculate the requirements" of each consumable. He remembers the long workdays well. "We had meetings that would end after 8:00 p.m. and that wasn't unusual."

In 1967, dire calculation results had bubbled up from mission planners to the top ranks of NASA. Worries about the lander's weight forced Marlowe Cassetti into special service. By this time, Cassetti, just like Pohl, led several organizational sections, and he supervised several dozen mission planners. "So, if you looked at all the numbers," he said, "it looked like we couldn't get there from here." The lander measured too heavy, the rockets too weak, and the subsequent requirements for fuel and liquid oxygen too overwhelming. "So we did something rather unique," he said. "We're talking about a time in computing history that was still [the] Stone Age. The late sixties were no better than the early sixties." Cassetti and his team needed to double-check every possible flight to the Moon and precisely report the degree of trouble—how many pounds overweight, how many ounces of fuel too short, and so on. And Cassetti started pulling at a thread that nagged him: hundreds of subcontracting companies had parts of Apollo to build, and each of them then passed a target weight number for their part to the main contracting companies, and the larger companies then passed *their* target weight for a whole space craft module on to NASA. As of 1967, it looked as though nobody could be absolutely sure of the weight of any of the three Apollo modules. Cassetti and his team would need to track down each exact weight like a retriever patrolling tall grass for birds.

Even with their new, sprawling center, mission planners still ran low on computing time. "One run a day is all you got," said Cassetti. "I happened to be over in another area of our building, and somebody had brought in a teletype machine. . . . I started looking over this thing and said, 'Boy what a great device.' . . . This was something that just absolutely blew me away." Within a month, he had one for his team. "It was a tabletop model," he said. The teletype used a revolutionary communications link, a modem. By nestling a phone's handset into the teletype, engineers could call and share computers, for instance, at the University of Houston (all at a dizzying thirty bits per second). This opened up the possibility of more than one run per day.

For these early computer users, glitches took on many forms. One particular project ran docking simulations routinely at midnight. Computers in one building controlled two model spacecraft bumping about in another building, but run after run came out garbled. After rounds of debugging, in both buildings, engineers finally located the problem somewhere in between: Late night janitors, innocently running a vacuum cleaner just inches over the sub-floor cables, created enough electrical noise to foul the signals.[15]

But with a new teletype machine up and running, Cassetti's team made progress. They inched closer to the elusive measure of Apollo's actual heft.

Whatever the weight of the craft nestled on top, was the Saturn V ready for actual liftoff? The year started with a definite no and moved to a strong maybe.

In late January, a test of the third stage ended in pyrotechnics in California. An explosion ripped the entire stage to threads of charred metal, while nearby buildings lost windows and even roofs—such was the power of hydrogen fuel. A supplier had provided a not-quite-right version of metal required for the welded seams. And the repeated rounds of testing had finally weakened one seam until it ruptured.[16]

When the Cape received new versions of all three stages in March, engineers found a total of 1,200 problems (many of them minor, like a bolt being slightly loose, but more than a thousand problems all the same). And while those engineers stacked the first Saturn V, others located new weld-line problems for stages still in production.

Technicians scurried about the stacked second stage and grimaced as they found those same problems on their rocket—it was time to take Saturn apart again.

In August, the restacked rocket looked more or less ready to go. The crawler carried forth a full Saturn V from the yawning assembly building. Slowly, tread by tread, the crawler took its balanced totem to the launch pad.[17]

Countdown starts, countdown stops, tests, alarms, pauses, much scratching of heads, and more starts and stops filled the following months. In October, problematic test results meant a new delay, and NASA slowly sucked all the fuel from the rocket. Some wondered if it would ever actually launch or whether it might just sit there as a sort of monument to hubris. Meanwhile, more and more Americans assumed the Soviets would reach the Moon first. Approaching the fiftieth anniversary of the October revolution, the world awaited a new dramatic achievement. And cosmonauts had started telling cheering Soviet crowds that when Americans finally got to the Moon, "we will be there to greet them."[18]

But a November window for the first Saturn V launch began to look promising, holding up against various minor delays and concerns. Finally, NASA would start using their mobile launch towers in succession. While one neared its launch date, the other two held rockets prepping for future dates with the heavens. Each tower could be moved about by one of the crawlers, with or without a Saturn V rocket. The towers included a fat, square base, where four enormous clamps held the rocket in place, and a tower of some 450 feet, topped by a crane and decorated with the nine swinging arms to nourish the rocket.[19] One engineer recalled standing near the top of the tower, as it embraced that first Saturn V rocket. He thought to himself, "No way in the world for that thing to lift off."[20] It was just too huge and contained too many systems, too many parts, wires, and seams. Even the more optimistic engineers had an attitude best expressed by one: "I hope it holds glued together long enough for us to get into orbit."[21] (See Figure 10.2.)

One of the last things engineers did before the approaching launch attempt was to carefully calibrate the rocket's sense of balance, a twenty-pound gyroscope spinning within the rocket's brain.[22] That brain, labeled simply the "instrument unit," sat near the top like a high-tech dog collar. It was just three feet thick, but fully twenty-five feet across, nestled between the top (third) stage of the Saturn V and its passenger, the Apollo spacecraft. The brain kept track of scores of sensors before and during a launch: temperatures, angles, flow rates, accelerations, and pressure readings, all streaming to and rendered by the brain. If the rocket started tilting out of line during flight, the instrument unit automatically ordered one or two of the F-1 engines to shift slightly this way or that to push it upright. The brain triggered the carefully set lines of explosives, separating the depleted stages one by one and letting them fall to

FIGURE 10.2 Aerial view of a fully stacked Saturn V and its launch tower moving toward the launch pad via a crawler hidden underneath. For scale, you may be able to make out a man in a white shirt and black pants walking across the platform to the right of the "USA" insignia. Apollo 15 is pictured here. (NASA photograph.)

the ocean. And, like a polite child taking a parent's car for a first drive, it sent rapid-fire updates to the engineers on the ground (see Figure 10.3).[23]

Meanwhile, America bit its lip and wrinkled its collective brow, but not anticipating a launch. The nation continued backing away from the promise of gleaming technology. Books offering readers mysticism, spirituality, or instructions for extra-sensory perception sold like never before. For every such purchased

FIGURE 10.3 The Saturn V's instrument unit on display in Huntsville. (Photograph by author and printed with permission of the United States Space and Rocket Center.)

book, a neighbor might pass it along. *I've had experiences like this—haven't you?* Some young Americans showed their rejection of early 1960s technocracy by embracing clothing referencing aboriginal Americans, with deerskin vests and fringe. NASA may have wanted to poke the night skies with rockets and capsules, but some of these hippies embraced "old ways" that would humbly leave the heavens alone.[24]

If Cape Canaveral engineers could have turned their attention from the mar-velous heights of their stacked rocket, they would have heard ever more divergent voices screaming about anything but Apollo. Race riots flared into 1967, turning es-pecially deadly in Detroit. And when fifty thousand antiwar protestors surrounded the Pentagon one October night, government troops chased them down, beat any resistors, and made mass arrests. Many Americans wondered if we shouldn't have our sights set firmly on earthly problems. Through no fault of their own, NASA's white-robed technicians swarming Saturn's launch tower were like so many priests on a modern Tower of Babel.

November saw the breathing of hundreds of engineers grow shallower. They were taking a risk with the first Saturn V. In order to speed things up, NASA's leadership had opted to skip testing each stage of the rocket individually. They'd try their first "all up" test on the most public stage, a sort of unpracticed high wire act that could end in a fireball. On the morning of November 9, they loaded it up: twenty-seven rail cars' worth of kerosene; twenty-eight trailer truck loads' worth of liquefied hydrogen; and finally, eighty-nine truckloads of liquid ox-ygen. NASA cleared all humans within a three-and-a-half-mile circle around the launch pad. (Snakes and sand fleas were on their own, as usual.) The capsule, for this launch, sat empty of astronauts. Engineers needed to see if this rocket, the most powerful ever built, could hold together, and then see if Apollo's command and service modules survived the launch and the rigors of space. Saturn's own brain would control the first minutes of the launch, with oversight from some 450 engineers in a "firing room." If they didn't like any of the signals coming from the rocket, or if it veered dangerously off course, they could abort the mission or even send a destruct signal to their beloved baby. Squinting from a remove of four miles were a press room and a viewing stand for dignitaries, NASA employees, and some NASA families.[25]

An intense fire marked the first many seconds of ignition. Smoke rushed from the pad as if it was possessed, but the assembled guests noted an oddly silent scene. Were they too far away to hear anything? The sound waves labored across the scrub land a full quarter-minute, but then they hit the crowds and buildings like a hammer. Spectators later shared a common experience, feeling vibrations yanking their sternums and ribs forward and backward, as if a rope connected them directly to the pulsing rocket. In the press room, Walter Cronkite, without breaking narration, watched engineers raise their arms and hold the windows overhead, trying to calm the undulating glass. "No one anticipated the true power of that rocket," engineer Chris Kraft wrote. "Ceiling tiles fell on television reporters at their anchor desks. The corrugated metal sheets covering the press viewing stands came loose and flapped dangerously."[26]

The first stage did its work, exhausting its fuel and oxygen in less than three minutes. The second stage, with five hydrogen-burning engines, took over and ran for the next six minutes. By the time it emptied and fell away, the remaining stage and its vacant Apollo ship were 118 miles above Earth.

"You could almost feel the will," one NASA official said later. Wernher von Braun shouted, "Go, baby, go!" Moments later he turned to a colleague, dumbfounded, and said he "never would have believed it possible." Indeed, as authors Murray and Cox later summarized it, "in its first trial, they had launched a rocket the size and weight of a navy destroyer, carrying eleven new engines [by stage, five plus five plus one] . . . new technology of all kinds, and had done it perfectly. There was simply no way to explain it."[27]

The launch was a stunning success. Not only had the incredible Saturn V aced its all-stages-at-once test, but the main Apollo modules had shrugged off the intense temperatures and vacuum in space. Compared to the crushing accident of January, the program's fortunes had turned completely. But the test still had its hiccups. Overlooked in the excitement, Saturn's main engines went into their guttering pogo like crazy—the old nemesis was back. The company building the engines admitted that, after years of full-time research, with an all-star team pulled from NASA, aerospace companies, and elite universities, they still failed to understand all the causes. A careless error also blemished the launch. An overeager controller working in the Australian station sent a command to the Apollo spacecraft which the on-board computer had already handled; in such a circumstance, the computer had orders to throw up its hands. *Okay, fine, humans. Do what you want.* It shut down its control of the engines, and by the time engineers figured out what had happened, the returning command module was moving a little too quickly. Impressively, a re-engaged computer made adjustments and brought the empty capsule down in the vast North Pacific Ocean, just two miles from its target coordinates.[28]

In the maiden voyage of the Saturn V, Wernher von Braun reached a sort of pinnacle for his long arc. Starting from the dreams of a kid, who had once strapped rocket motors to his wagon, to building a feared weapon for Nazi Germany and then hustling parts of it to America, he'd always hoped to build a rocket powerful enough to lift a real spaceship.

He returned to Huntsville, a town that had truly become home to so many Germans. Thanksgiving that year brought the successful engineers to their families, for at least a day. The 1967 *Huntsville Heritage Cookbook* compiled favorite recipes from local housewives. Mrs. Wernher von Braun led off the "Foreign Foods" section, setting down her nut cake recipe, with pecans and lemon icing, and her "Food of the Gods," a rum-soaked whipped cream with chocolate shavings. Whether or not the von Braun family passed these around their Thanksgiving table, Wernher faced very earthly and mortal matters back at work. On November 29, he gathered his adoring troops in an auditorium. Far from taking a victory lap, he relayed the sorry news: given a new set of NASA budget cuts, he would be forced to eliminate seven hundred positions by year's end.[29]

11

1968—OF TIMELESS VIEWS AND NEW PERSPECTIVES

Many Apollo engineers admit they didn't see the late 1960s coming, and they hardly experienced these turbulent years. "I watched no television, read no newspapers, came to work at six in the morning and worked until nightfall, six or seven days a week for years," said one engineer, voicing the memories of many. Yet, their spouses, their parents, and, for some, their kids had to be worrying aloud. My father recalls requesting a draft furlough for one of his young engineers, and my parents together describe stomachs knotted while helplessly hearing the news in these years: mass protests, riots, protests, and sad lists of soldiers. Henry Pohl's son Karl recalls Vietnam updates filling the early calm of the family home. "Every morning my mother would have the radio on as she was making breakfast," he says. "Every morning, to hear the body count."[1]

In mid-March, U.S. deaths in Vietnam surpassed five hundred in a single week, and the nightly news announced that Vietnam had now moved past the Korean conflict in American losses. The year would be the deadliest of the war. As the nation wobbled its way toward an ominous-feeling presidential election, radical activist Eldridge Cleaver wrote from prison that America was surely scaring the people of the world. "They must feel like passengers in a supersonic jet liner who are forced to watch helplessly while a passel of drunks, hypes, freaks, and madmen fight for the controls and the pilot's seat."[2]

By this time, Martin Luther King Jr. had turned his nonviolent focus to poverty, but his efforts felt increasingly futile. Associates later recalled him falling into a depression by 1968. "Maybe we just have to admit that the day of violence is here," he told one close friend. "And maybe we have to just give up and let violence take its course." King then watched an attempted peaceful Memphis march fall to chaos on March 28, with protestors battling police and shattering store windows. In early April, he began to prepare a somber sermon. As Apollo engineers thought only of taking the nation upward and outward, King titled his draft "America May Go to Hell."

In the midst of these anxieties and despite ebbing enthusiasm for technology, space still offered Americans a fascinating escape. In 1968, many more tourists visited NASA's Manned Spaceflight Center in Houston than the Grand Canyon.[3] But outside the agency, the nation's space fascination shifted from bold and exploratory to neurotic and self-doubting. Perhaps, the culture seemed to suggest, we were not worthy of the cosmos.

The strikingly realistic film *2001: A Space Odyssey* debuted that spring. (Director Stanley Kubrick had benefited from the early space industry layoffs, hiring newly migrant engineers as consultants.) *2001* featured many parts of von Braun's long-time playbook: an orbiting, spinning space station; a well-populated human presence on the Moon; routine and commercial-grade spaceflight; and the ability to travel to other *planets*, including mysterious Jupiter, beckoning from nearly half a billion miles away. All this by the year 2001? Audiences could lean forward in their seats and nod their heads, given the pace of the 1960s.

Despite the realism of technology in *2001*, the mood was not as optimistic and outward-focused as NASA may have hoped. In the film's central premise, space had actually come to Earth first, instead of vice versa. A creepy and foreboding alien intelligence had some sort of plan for the inferior, ape-like humans. Highlighting a dread for the unknown universe, the movie scored alien technology with an unnerving György Ligeti "sound mass," with a choir voicing hellish cacophony.

Another 1968 film, *Night of the Living Dead*, envisioned a terrifying NASA blunder, as an otherwise happy probe, returning from Venus, radiated Earth and—whoops—launched a zombie apocalypse. In another writer's mind, a probe would bring back a horrific disease, clotting the blood of any human it encountered; in 1968, Michael Crichton busily penned *The Andromeda Strain*.[4]

Turning reflective, the popular imagination seemed to ask, what did space mean for earthly life? The year witnessed the publication of *Chariots of the Gods*, promoting a theory (not unlike *2001*) that ancient alien civilizations had visited Earth and assisted our hopeless, knuckle-dragging ancestors. Reversing a von Braun–type aspiration, this new view downgraded humanity. Maybe space was too big, too old, and too smart for us.[5]

As NASA prepared the second launch of a Saturn V rocket and knocked the kinks out of the lunar lander, America marched toward what became the "summer of hate." President Johnson's most applauded moment in his January State of the Union address was his condemning of urban violence. Instead of another pillar of the Great Society, he announced a "Safe Streets" Act.

He was also immersed in budget worries, watching the Revenue and Expenditure Control Act, including billions in budget cuts, make its way through Congress. The entire globe had financial jitters that spring. A spike in gold speculation led to an emergency decision to shutter the London gold exchange in mid-March. Western leaders hurriedly came together in Washington, D.C., trying to avoid what could become a global financial collapse.

At the end of March, the president, a steady political patron for space exploration, made a stunning announcement: he would not seek reelection. So, *if* NASA fulfilled Kennedy's promise, it would happen under new leadership, for a president who'd played no part in the initial challenge.

Meanwhile, the Soviets announced a Moon-worthy feat, taking a new, heavier vehicle—unmanned but suitable for a cosmonaut—to a Moon-like distance and back to Earth. The Zond 4, at nearly the same weight as Apollo's command module, traveled 220,000 miles away, stayed free of cabin leaks, and returned to Earth.[i] They were still in the race, with more missions on deck.[6]

The engineers collectively swallowed, blinked, and returned to preparing the second launch of a Saturn V. They worked through every detail to better Apollo's odds and minimize risks—for astronauts and spectators alike. That spring, engineers decided to change the rocket's ascent path. It would head eastward and out over the ocean earlier. This so-called "range safety" decision would best protect the towns of central Florida from any mishap in the first few minutes.

The lunar lander continued to suffer an awkward adolescence. Its thin, weight-saving wiring would often snap. Managers begged lander workers to be as careful and gentle as possible with their movements and even their footfalls in the fragile craft. At the same time, NASA still worried that the lander might be too heavy. As Cassetti and his group continued fine-tuning their calculations, NASA offered the lander-building contractor financial incentives to shed further pounds.

On April 4, the second unmanned Saturn V rumbled upward, rippling the sands of the Cape. Problems started right away, but not for the engineers at the Cape or in Houston. The Huntsville team received the data collected by the rocket's instrument unit, and they got an eyeful during stage I. The ship reported a terrible pogo effect, with vibrations surging up and down the rocket as if it were backfiring. The chugging yanked the structure back and forth with a force ten times greater than Earth's gravity itself. It lasted for about the first ten seconds and then went away. The engineers exhaled. But when they compared notes with their Houston counterparts, they learned that any pogo that violent would have severely injured human passengers. Even if astronauts could withstand such shaking, the vibrations surged past the Apollo spacecraft's design limits, risking ruptures, breaks, or leaks.

The Saturn survived its first few minutes, discarded its troubled first stage on time, and fired the second stage engines, with five hydrogen burners leaving a fluffy white

---

[i] The Zond 4 actually went in the opposite direction of the Moon, probably to avoid worrying about the Moon's gravity. The Soviets aimed to recover it on Earth, but after a guidance malfunction, they gave it a self-destruct order during re-entry.

trail of water vapor. After four minutes, one of the engines turned itself off. A second later, another did the same, as if joining a sit-in. The rocket limped along with only 60 percent of its designed thrust. Engineers had thought the odds of losing two of these engines in one launch to be statistically equivalent to zero. With two side-by-side engines out of commission, the other three now provided unbalanced thrust, and the rocket risked tumbling out of control. The Saturn's brain fought to keep the rocket on course by pivoting the three working rockets. It wasn't a pretty path—engineers watched telemetry data plotting zigzags across the sky—but incredibly, it maintained control.[7]

By asking the third stage to do a little extra work, engineers could still salvage the mission and put the Apollo capsule through its paces.[ii] Burning this stage longer than normal put the third stage plus its Apollo craft in orbit—not the originally planned orbit but good enough for now. Then, a few hours and two orbits later, engineers moved to their next test: They would restart the third stage engine, simulating Apollo's eventual push away from Earth and toward the Moon. Like returning to a warm car, key in hand, they sat down and cranked the ignition. Nothing—not even a chug or a puff. The engine would not restart. If the mission had contained astronauts and a lunar lander, they would have been all dressed up with nowhere to go.

Finally, given all the engine trouble, the empty Apollo capsule came in more slowly than its guidance computer expected. A software bug joined in to torment the mission. The mistaken computer code would never have affected a mission moving at proper speeds, but here, it splashed the capsule down fifty miles off target. The extra time for the capsule floating at sea emphasized yet another problem. The three-seater capsule sported a larger profile than its predecessors and with a relatively light weight, it made for an excellent sailboat. Resting on an inflatable collar, it had very little of itself below the water line, and the Apollo capsule ran with the brisk Pacific breeze. A recovery helicopter chased it down and dropped the requisite team of Navy Seals. They planned to climb onto the capsule and affix a cable, but the capsule sailed faster than the Navy Seals could swim. In later missions, the recovery team learned to take a good wind measurement and then drop the Seals in front of the capsule's nautical tack.[8]

The mission left the engineers with dry mouths and knots in their stomachs: a first stage that would shake the fillings from an astronaut's teeth, and then *three* separate engine failures in the upper stages. But as NASA officials prepared to brief

[ii] In one of those names that can only come from scientists, engineers, or rock bands, the third stage was actually called "S-IV," because, in earlier planning phases, they considered a total of four stages, but they eliminated what would have been S-III. And, while we're speaking of numbers, the Saturn "five," did not relate to stages or number of engines. It was simply the fifth milepost in that rocket series.

reporters with the flight's bad news, Apollo fell quickly and quietly to the back page of America's concern by the end of the day.

Setting aside his new sermon late on the same afternoon, Dr. King had stepped outside his Memphis hotel room. Across the street, a fugitive convict named James Earl Ray watched King from the bathroom window of a boarding house. Earlier that year, the bigoted Ray had volunteered his time to the segregationist campaign of George Wallace. He raised a rifle to the window sill and felled Martin Luther King Jr., a man not yet forty years of age. Engineers may have looked up from their work or fielded worried calls from home, but they did not stop.

"For those of us involved in getting to the Moon, it was like we were on this little island doing our thing," engineer Glynn Lunney said later. Raised in a Pennsylvania coal town, Lunney had joined NASA just months after graduating from college and had worked through each step of the space age, eventually as a flight director. "It's shameful to say that, but it's true. We were so consumed by getting to the Moon, we had little time to pay attention to anything else." Amid national if not universal grief, engineers continued finding, fixing, and improving one technical issue after another.[9]

Aside from the all-consuming hours, the Apollo engineers had an understandable human choice in front of them, every day. On the one hand, America presented a dizzying set of problems that would not necessarily yield an inch to determined logic or good intentions. On the other hand, America presented them a single problem—a daunting one, a dangerous and expensive one, absolutely—but a problem on which they found handholds, could clamber to progress, and just maybe solve. Some have admitted they had doubts, even then, about the money flowing around and over their desks, given all the problems in America. But they all wanted to finish this gargantuan job. Order, structure, selfless teamwork, and aspiration contrasted completely with the teetering world beyond their offices and laboratories.

To rub a bit of salt on NASA's new technical problems, the Soviets put their Luna 14 probe into orbit around the Moon on April 10, and they prepared to unveil a new, more powerful rocket booster, something that could start a manned Moon mission.

Then NASA nearly lost another astronaut as he trained for a lunar landing. To give astronauts a feel for an otherworldly flight in the strange lander, engineers had improvised an ungainly, hovering machine. Operating at a military base near Houston, it used a jet turbine to fight off about 85 percent of Earth's gravity, letting the skeletal craft more or less experience a Moon's worth of gravity in what was left. This contraption, dubbed the "flying bedstead," malfunctioned in May. During a practice landing, it began leaking fuel and suddenly all its controls went dead. As it plummeted to the ground, the astronaut pilot, Neil Armstrong, ejected just a fraction of a second before losing his life. He maintained enthusiasm for the bedstead,

FIGURE 11.1 The Lunar Lander Research Vehicle, aka "the flying bedstead," during tests at NASA's Flight Research Center in 1967. (NASA photograph.)

however, and later claimed it provided the best practice for landing on the Moon.[10] (See Figure 11.1.)

Meanwhile, NASA quietly started racking up a string of successes. By summertime, managers felt hope returning as determined engineers felled several major problems.

Finding and installing a pogo fix in the first stage consumed over 30,000 manhours packed into just a couple of months. The whole assembly had resonated like a demonic organ pipe once the engines had started sputtering. The solution was ultimately a familiar one for the Huntsville team. The engines had vibrated about 5.5 times per second, while the rocket's liquid oxygen pipes naturally shook at a toosimilar 5.25 times per second.[iii] Engineers had to change one or the other, so that the two components couldn't talk so easily and reinforce one another. Once again, they wouldn't try to eliminate the mercurial pogo, but they worked to keep it from spreading around the rocket. Given the lack of time and declining budgets, they ruled out making significant changes to the pipes themselves—that would take too

---

[iii] These frequencies are too low for human hearing, but they might coax a reaction from a rhinoceros or a humpback whale.

long and eat too much money. Instead, they added a little helium gas to the flow of liquid oxygen. (The helium was already available, since the Saturn used it to push kerosene from its fuel tanks.) Mixing helium with oxygen didn't affect engine performance, but it dropped their vibrations from over five times per second to just about two times per second.

The three remaining engine failures (the two in the second stage, and the solitary one that wouldn't restart in the third stage) were all in the hydrogen burners. Luckily for engineers, the first engine that failed (in the second stage) and the one engine that wouldn't restart in the third stage showed exactly the same problem. In each case, a steel fuel line, barely half an inch across and six feet long, had ruptured, leading the engine to automatically go dormant to avoid an explosion. Both lines had ruptured in the same spot: a little accordion-like "bellows" intended to help it withstand vibrations. For a shaking line, whether from unwanted pogo or just from the violence of a launch, engineers had figured flexibility to be a plus. But why would these lines break now, when they had shown no hints of trouble during many Earth-bound tests? Detective work showed an unlikely culprit: frost. In all engine tests on Earth, natural humidity frosted the outside of the frigid line carrying liquefied hydrogen. And, in a surprise to engineers, the frost actually formed an extra layer of protection for the relatively weak bellows. During a real launch, however, the rocket had already climbed past most of the atmosphere when these engines needed to fire. With no available water vapor, no frost formed. Without frost, the jostling lines were free to snap on their way to space. Engineers revamped the fuel lines and eliminated the bellows.

But the third failed engine, the one that cut out right after its neighbor, hadn't suffered a fuel line break. Here, engineers traced the fault back to a wiring mistake connecting second-stage engines. When the first engine suffered a busted fuel line, the faulty wiring took the neighboring engine down with it.[11]

By summer, von Braun, his team, and their various contractors had the engine fixes in hand. But the decade had dwindled to just eighteen short months. They had not yet put a human being on top of the mighty Saturn V, after one good launch and one nearly disastrous one. Their president and most important supporter would be leaving office, and their funding had already started to wane.

In October, a reporter asked von Braun what he was most focused on during the home stretch. "It may surprise you to hear this," he said, "but for the last two years my main effort . . . has been following orders to scrub the industrial structure that we had built up at great expense to the taxpayer." The once-playful rocket scientist grew more direct with the press and more frustrated with those calling the shots. Whether or not many Americans noticed the shift from bright-eyed, do-everything von Braun to a more bitter version, his supervisors at NASA headquarters noticed.

And he wasn't done with the reporter's October question. "The sole purpose," he continued, "seems to be to make certain that in 1972 nothing of our capability is left." He sensed America finishing its long and expensive honeymoon with space.[12]

Even without these headwinds, Apollo still faced two significant problems: whether all the pounds added up to an achievable Moon journey, and whether they would ever have a vehicle to land there. In early summer, the latest lander slunk into Florida's Kennedy Space Center, where inspectors found a bunch of the usual suspects, counting more than one hundred flaws. Windows failed under a vacuum test. Some wires had snapped during the trip from Grumman's manufacturing plant, and many thin metal elements showed cracks again. Everyone now recognized that the lander wouldn't debut until 1969 if it ever worked at all.[13]

Meanwhile, Marlowe Cassetti continued his quest to learn Apollo's total weight, lander included. He and his staff leaned on their new teletype machine, regularly using its phone connection to the computers at the University of Houston. Having ferreted out the actual weights of the rocket, stage by stage, and each of the modules, his team computed potential flights to the Moon. And in the spring of 1968 they could see the finish line. "One of the guys that worked for me and I came in one day to finish up our analysis," he said. "And we came in there and we found out it was Memorial Day and the whole place was shut down. We had no power in our building." It seemed like they would *have* to take the weekend off.

"Wait a minute," he thought. "I've got a telephone at home. I've got electrical power." Why not drag the teletype home for the long weekend? The two engineers loaded the bulky beige machine into his car. "It weighs a ton," he said. Cassetti had to sweet-talk the guards: He wasn't trying to steal government property. "Took it home, cleaned off my dinette table, which was the closest to the [cord-bound] telephone and had this thing running, and we worked there all day until we got off our final results."

What did they find? Was NASA a go or a no-go for the Moon?

"The results basically were, we don't have a problem." The alarm had sounded in 1967, as the whole enterprise looked too heavy to approach the Moon, but Cassetti found a psychological cause instead of a technical one. As good engineers will do, each sub-contractor, contractor, or NASA group wanted to give themselves just a tiny bit of wiggle room, in case they needed to add one extra bolt, one extra chip, or a little extra sealant to something heading to space. With all the conservative fudge factors added together, Apollo looked too heavy on paper, but once Cassetti and his crew extracted the actual up-to-date numbers, sans fudge, the total weight looked perfectly liftable.[14]

More good news arrived in short order. The main Apollo craft, including the crew capsule and the service module that followed it, responded well to all recent tests. The new designs, emerging after the tragic fire of 1967, set a record for fewest problems on arrival. NASA's unusually brutal oversight of the contractor's work had paid off.[15]

Spacecraft inspections created a new type of headache for Henry Pohl at about this time. Among his many responsibilities, he now oversaw explosives on Apollo—the planned and helpful kind. The modular spacecraft had so-called pyrotechnics, or low-power explosives used to jettison pieces or separate modules from one another at critical moments. "We took a neutron radiograph of all that hardware . . . and the explosive would show up as a black or dark image," he said. With his more experienced men unavailable for a review meeting, he sent one of his newest employees along to examine the latest radiograph images, not expecting anything troublesome. "I just wanted the guy to have the experience," Pohl said. "He hadn't been with us very long."

But in looking at the images, the employee was puzzled. "What are all of these little spaces in here, these bright spots in here?" he asked Pohl. They soon realized he was looking at gaps in the explosive ring. "And some of them were a quarter inch or longer," Pohl recalled. "Then I came unglued. I mean, we're getting ready for a launch, and you obviously can't fly that junk." To work properly, the explosive needed to form a continuous line all the way around the junction between, for instance, the command module and the service module. If the explosive charge failed to cleanly cleave the modules before re-entry, the astronauts could be stuck with dangerously tangled craft at the worst time, heading into Earth's atmosphere. At more than twenty thousand miles per hour, the air resistance could rapidly spin the conjoined modules into a deadly tumble, and even if they stabilized, the heat shield would be stuck between the modules, unable to serve its protective purpose.

"How did [it] get by?" Pohl asked. "There were twelve or thirteen people that had signed off." A number of people at different ranks had cleared each radiograph scan. Pohl started calling the signatories, one by one in reverse order. Each had a story. "Well, I don't know anything about N-rays," said one. "But we get this certification from the company that takes the pictures." Finally, at the end of this chain, Pohl interrogated the fellow who applied the first signature. "I personally look at every single N-ray that goes out of here," the man proudly assured Pohl. "And I make sure that it is a good N-ray."

"What do you look for?" Pohl asked.

"Well, first thing I do," the man said. "I look to see if the film is fogged." Next, he made sure the images were in focus.

"Well, what about the hardware that you're taking pictures of?"

"I don't have the foggiest idea of what we take a picture of."

So, the first technician certified that it was a good and proper image of the subject, with no clue what the Apollo equipment should look like. Was it in focus? Showed no artifacts or smudges? It was good to go. But the picture then migrated through a growing bureaucracy. The meaning of the signatures quietly morphed from "the image is in focus" to "this explosive ring is in perfect shape for its job." They had nearly launched a flawed craft, based on sloppy semantics and the seductive comfort of administrative bloat.[16]

In late July, three main drivers of Apollo gathered in a Houston office. Each of the three could trace his roots back to the headwaters of NASA and was now an "old man" by the agency's standards. George Low, a mild-mannered, well-regarded engineer and manager, had just returned somewhat dejected from the Cape, where he'd seen the hot mess of the lander up close. He called center director Bob Gilruth and the czar of Mission Control Chris Kraft to his office.

Low had a bold idea that he wanted to bounce off his two most trusted colleagues before it went any further: What if they shuffled the planned missions and really went for broke? They'd fixed the Saturn V, and two of the three Apollo modules looked solid. So, Low suggested, just forget the lander for now. They could push a mission all the way to the Moon and back. He wasn't suggesting a fancy flight—just a big figure eight, like ice-skating out into the void, gracefully dipping behind the moon, and gliding serenely back to Earth. They'd make no landing but, with luck, they could notch a major milestone before the Russians sent cosmonauts to do it in their Zond.

His colleagues both liked it. They checked the idea in a carefully curated ring of personnel. One person worried that their model for the Moon's gravity might be too naive. The lumpy Moon could cause problems as it suddenly changed its pull on a craft. Low's idea morphed. If they were flying all the way to the Moon, they may as well cozy up to it, take some laps, and measure its gravity more precisely. The new mission might not only prove they could fly a 480,000-mile roundtrip, it could also gather crucial information for future, landing-focused missions.[17]

Engineers in the planning program recall the various stages of an expanding ripple. There were just fifteen engineers in one of the early meetings. "This was of course hush hush. . . . It was a pretty scary flight," engineer Mac Henderson said. "The first time we took a crew beyond Earth orbit."[18]

A small group flew from Houston to Huntsville in early August and consulted with Wernher von Braun and his top brass. After a week of thinking it over and consulting his team, von Braun voiced strong support for the circumlunar voyage. To him, the decision boiled down to whether or not the Saturn V was safe enough for a crew. "Once you decided to man [the rocket]," he said. "It did not matter how far you went."[19] By August 14, strong consensus spanned the leadership in Houston, Huntsville, and the Cape: Not only could this mission work, but it was an obvious choice, especially given reports of the Soviet's progress with their heavy Zond craft.

Privately, NASA decided that the eighth Apollo mission, the first putting men atop a Saturn V, would head to the Moon. Most NASA employees, to say nothing of the contractors, knew nothing about it at this point. For months to come, NASA informed the press and public of the standard plan: Apollo 8 would stay in Earth orbit, testing docking maneuvers between the command module and the lander. "We were in a situation where we couldn't tell our colleagues what we were working on," mission planner Lee Norbraten said, "much less our families."[20]

Frank Hughes, the simulations expert, recalls enjoying a rare day off that fall. "I was down at the pool. It's Florida. It's Saturday morning, looking forward to the three-day weekend," he recalled. "I walked back [inside] to get Coke or a beer. . . . The phone's ringing, I pick it up." It was his boss, bluntly telling Frank that they were sending the eighth mission to the Moon. "This is Labor Day, and we're going on December 21st," he says, still expressing disbelief. He knew he'd spent his last pool-side moments for the year.

Hughes immediately flew to Houston and found himself in a crowded, smoky conference room. The meetings in Building 30, home to the Mission Planning and Analysis division, lasted through the holiday weekend. The assembled quizzed Hughes about the status of the simulators, given the enormous shift in plans. "We'll be ready," he said at the time. In hindsight, he says, "I had no clue what the hell I was doing—how were we going to get all that done?"[21]

One of the most significant challenges involved speed. Moving faster than any human had moved wasn't a problem by itself. Any little asteroid or a solar flare would pose the same risk to Apollo. The problem arose when the craft came back to Earth's atmosphere moving a good deal faster than any previous mission. Nosing into the air with extra speed meant the craft would generate that much extra heat during braking.

With their "translunar" scheme still a secret, the engineers first had to hear of another step of Soviet progress. A Zond craft—still unmanned but plenty roomy and holding a breathable cabin atmosphere—went all the way around the Moon, snapped a picture of Earth, and survived a splash down in the Indian Ocean. How long would it be until they tried this with a cosmonaut on board?[22]

Before any public announcement of NASA's new plan, the leadership, going all the way to President Johnson, wanted one more successful mission under their belts. Apollo 7 launched in October. Though it did *not* use a Saturn V, this mission kicked the tires for the command and service modules, leaving them up and toiling in Earth orbit for nearly eleven days. This was America's first three-person mission, and the first inhabited test of the Apollo spacecraft. Though the crew became unusually grumpy as the endurance test wore on, they caught a surreal new sight. The Apollo craft routinely dumped excess water and waste, forming glittering fields of floating ice crystals and snowflakes, particularly visible during orbital sunrise or sunset. An astronaut reported seeing a perfect silhouette of their ship, the pointy command module atop the soup can of the service module, as a shadow set against a drifting swarm of ice particles.

A sort of spacecraft burping provided another surprise for the mission. One of the lesser known tasks of the mission involved Apollo flying about one hundred miles away from its original orbit, where the third stage of the rocket still floated as an inert shell. Engineers wanted to practice closing on a target from a great distance, hoping to gain practical experience with the tricky game of rendezvous. When Apollo returned to the third stage, engineers were surprised to see it had changed its orbit from the day before. In time, engineers determined that the stage belched fuel vapor on occasion, with each burp giving it a little propulsive kick. As the stage floated from the night-side of Earth into sunlight, the sudden warmth vaporized some of its un- used rocket fuel, leading to the orbital dyspepsia. The Apollo program slowly sorted out these little kinks, mission by mission—every little bit expelled, from fuel burps to urine dumps, introduced an effect like a tiny rocket engine.[23]

With the success of the first manned Apollo mission behind them, NASA made the public announcement in November: in one month, our species would try for the Moon. CBS television had just moved from thirty minutes of nightly news to a full hour, and they now included frequent updates for the upcoming Apollo 8 flight. Against 1968's backdrop of the Vietnam War, student protests, Soviet tanks rolling into Czechoslovakia, and a chaotic political convention in Chicago, a group of plucky engineers said they would, at least for a few days, try to transcend it all.

But now NASA needed an extra favor. Typically, an Air Force general stationed at Cape Canaveral coordinated with the navy for all capsule recovery operations, but not this time. All special requests for the Pacific Fleet needed to go directly to Navy Admiral John McCain, whose son (an eventual U.S. Senator from Arizona) languished in a Vietnamese prison camp. Local military officials in Florida told NASA they weren't willing to ask the Admiral to volunteer part of his fleet over the Christmas holidays, so engineer Chris Kraft flew to Honolulu to present the plan and beg for help.

He awaited his chance to tell Admiral McCain the plan, sitting in front of an am- phitheater that filled with more and more navy brass. "At 10:30 a.m. sharp," Kraft later wrote, "someone yelled, 'Attention!'" and McCain strode in.

"Okay, young man," he said to Kraft. "What have you got to say?" Kraft made his pitch and then asked for the navy's support throughout the Christmas season, to re- cover, hopefully, a live crew that had flown to the Moon and back. Admiral McCain thought for a moment, working on a substantial cigar. "All of a sudden, he stood up and threw it down on the table," Kraft wrote.

"Best damn briefing I've ever had," McCain said, at substantial volume. "Give this young man anything he wants."[24]

In November, the Soviet Union swung another unmanned Zond capsule around the Moon, much closer this time, and it survived a return to Earth, smacking the ground in Kazakhstan. Cosmonauts had been eager to climb aboard and be the first to leave the confines of Earth orbit, but each of the returning capsules concerned Soviet engineers. The one in September had entered the atmosphere at the wrong angle, and the g-forces would have killed any crew inside. Then the more recent one had literally blown a gasket and expelled its air, killing a few animals on board. Nonetheless, according to some accounts, a cosmonaut did board a Zond-loaded rocket and prepare for launch in early December. Engineers detected some worrisome measurements from their rocket's first stage, however, and they coaxed him out of the ship and away from the pad. When they went ahead with an unmanned launch, aiming again to loop the Moon, the mission exploded within its first minute of flight.[25]

By late December, everything aligned for NASA. While planning a mission usually consumed a busy year, the engineers had crammed it all into four months. The world's most powerful rocket, making only its third launch, would take three humans up and away from Earth. If all went as planned, it would set distance and speed records for the species. It would also take us fully away from the protective confines of Earth's magnetic shield for the first time. (Our planet's magnetic field deflects otherwise dangerous charged particles streaming in from the sun and the cosmos.)

The Saturn V's launch on the morning of December 21 rattled the astronauts to their bones. One called it "an old freight train going down a bad track." As the rocket's brain made little corrections, pointing its massive engine bells a few degrees in one direction and then the other, it pitched the astronauts side to side in their harnesses. One said he felt like he was riding the tip of a car's radio antenna. And it was incredibly loud, like no other rocket. The astronauts literally could not hear one another.

When the first stage cut out, the force pushing them upward suddenly disappeared. The astronauts felt themselves at the peak of the world's largest roller coaster, awaiting whatever came next. One said he felt sure he'd lurch forward into his instrument panel—he held his arms up to protect himself. But the second stage kicked in, and as he flew back into his body-molded seat, he smacked his own helmet with his suit's heavy metallic wrist. After a little minor pogo in the second stage, the ride ended in a perfect Earth orbit. Everything looked good to go.[26]

Just ten years earlier, even with the inception of NASA, the next step must have been inconceivable. Apollo, still connected to the top (i.e., the third) stage of the Saturn V, restarted the single hydrogen-burning engine. The mission sped from what had become a routine seventeen thousand miles per hour in orbit, to roughly twenty-five thousand miles per hour, faster than any humans had ever moved with respect to terra firma. As the Apollo craft leapt away from the security of Earth, the engineers in Houston paused from their obsessive monitoring. "There was nobody even breathing hardly," according to Mission Control flight director Gerry Griffin. "And it was almost like a religious experience." Most of the engineers felt they'd won the space race at about 10:00 a.m. on December 21, 1968. "We were headed out," Griffin said. "We all kind of looked at each other and said, 'Well, we've done it now.'"

With Apollo 8, humanity entered a sort of three-dimensional existence in the cosmos. For tens of thousands of years, humans occupied a thin sheet of comfort enveloping our planet. Even including a person in Earth orbit, just 100 or 150 miles up, our existence had been confined, by analogy, to the thickness of the skin surrounding an apple. If we could ignore gravity, a humble *car* could cover the distance between the ground and Earth orbit in under two hours, like a trip to grandma's house. In 1968, we leapt up and away from that thin layer of our history. Human existence no longer mapped to a flat projection of the globe, and we could no longer track our every migration and milestone using latitude and longitude.

"You're really happy," Griffin recalled, "and then all of a sudden you say, 'Good gosh, I got to get this thing back on the ground again.'" With the spell broken, engineers checked the trajectory, and it looked perfect. In fact, the initial path was so perfect that Apollo 8 skipped the majority of seven pre-planned mid-flight adjustments; at most check-in points, the astronauts found the ship not a single degree off course. All the monitored values—pressures, temperatures, rotations—were good. How about the vital signs of the human payload? One of the astronauts wasn't doing well—he felt weak and started vomiting.[iv] "Our first thought was that there's something we don't understand about going toward the Moon," Griffin recalled. "It's going to make them all sick. And we got a disaster on our hands."[27]

Outside the protection of Earth's magnetic field, could there be some sort of space radiation sickening the crew? Or could the astronaut have caught the nasty Hong Kong influenza running around the globe that winter? NASA had tried to quarantine the astronauts in advance, but Johnson, wanting to enjoy the last days of a difficult presidential year, waved his hand at the doctors and had the astronauts attend a December banquet with him.

Before Earth-based doctors could fully panic, the first astronaut stopped vomiting, and the other two never started, despite sharing a small volume of air with their colleague.

[iv] Astronaut Frank Borman.

In time, NASA learned that this was just a normal bout of space sickness that would affect many astronauts once they unstrapped themselves from their seats. Simulations expert Frank Hughes describes it as simply having too much room in space for one's own good. In the first phases of manned spaceflight, "the craft was so tiny, nobody got sick," he says. But now, with a three-seater and extra room, "it's the combination of being able to get out of your seat and float." Floating with no real up or down, and no seat, and no forward to face, has made many astronauts ill over the years.[28]

Arguably, the intensely macho tradition of test pilots, and then test pilots who became astronauts, slowed NASA's progress in understanding space sickness. Trained for extreme motion, and coming from a culture of absolute toughness, astronauts didn't want to admit any discomfort, never mind discuss any details with a doctor. An astronaut on the following mission, Apollo 9, became very ill,[v] and his post-mission efforts to research the problem—including an amusement park's worth of nauseating rides attempting to induce the feeling again—contributed, in some accounts, to his fall from the astronaut brotherhood.[29]

Now on its way to the Moon, a ship for the first time had to deal with intense sunlight, with no help from even the outer wisps of an atmosphere, and with no periodic shade from Earth. It would suffer nearly three straight days of 250° Fahrenheit on one side of the ship, and –250° Fahrenheit on the opposite side.

Some readers may already be wondering why it took so long to reach the Moon. If it is about two hundred fifty thousand miles from Earth, and if the Apollo 8 mission revved to twenty-five thousand miles per hour, then, just like calculating the time for a train to reach a station, it should only have consumed about ten hours. But Earth's gravity doesn't quit. Its gravitational pull continues without boundary, just with a weaker effect at greater distance. Hence, Apollo slowed all the way to two thousand miles per hour during its outward journey, as Earth tried pulling it back. Moreover, no Moon mission traveled on a straight line. Apollo's path was more of a gentle S shape, starting "behind" Earth, arcing out in front of the moving Moon, and then gently sliding in behind it.[30]

To handle the temperature extremes, engineers had originally planned a special type of sun shield for this leg of the trip, but one engineer came up with a more cost-effective solution. "We called it barbecue mode," my father says with a smile. By slowly rotating the entire craft, spinning one full revolution every hour, engineers ensured that no one part of the hull got too hot or too cold. One might then logically assume that the Apollo missions flew like a nicely thrown football, spinning steadily and aimed at the Moon. But the heating in that case would still be uneven. As the

---

[v] Russell "Rusty" Schweickart.

craft chased the Moon, its nose or its rear end would have taken more than its share of sun, spinning or no. So instead, after an initial push toward the Moon, each mission pivoted its nose "upward," facing the North Star, and spun toward the Moon like a child's top gliding across a well-waxed floor.[31]

Barbecue mode created an unexpected noise problem during what should have been a serene trip. A child's top keeps a steady spin, in part, because it is perfectly balanced around its centerline. But if you filled the top with ever-changing fluid levels, and astronauts moving within it, the top would not maintain its spinning axis. It would wobble. The Apollo guidance computer fought this effect, firing Henry Pohl's little thrusters to keep things spinning perfectly around the original centerline. Engineers prefer a tidy system after all. The thrusters fired in brief bursts, registering each time as a loud bang inside and making sleep nearly impossible for the astronauts. (Eventually, some physicist types within NASA convinced the engineering types that it was harmless to let the craft "find its own axis" during barbecue mode—Apollo didn't need to maintain a flawless spin, and future astronauts got some rest.)[32]

Engineers monitoring the flight recall a magical Houston moment for the outbound trip that December. As with my father's earlier, math-choked research paper, Apollo employed a patchwork path to the Moon. For most of the trip, NASA referenced the craft's position to the center of Earth, but once the Moon had more of a gravitational say, engineers switched to a new, lunar-based referencing scheme. This time, the engineers weren't just dealing with equations; now a slip-up could lose three lives. "The onboard computer for the command module would do this big switch, would recalculate everything and switch to a lunar-centered coordinate system," Apollo computer guru Jack Garman said. Having recently turned twenty-four years old, Garman oversaw a Mission Control console for Apollo 8 with some other young colleagues. A small indicator was supposed to light up when Apollo's onboard computer made the mathematical transition from Earth-centered to Moon-centered reality. "At two o'clock in the morning or something on the way to the Moon," Garman recalled, "you have nothing to do. Long moments of boredom waiting for something horrible to go wrong. So we were guessing the exact point at which the light would come on." They essentially took bets, down to two-second intervals. "Yes, we're a little nuts, but what do you do?"

When the indicator silently popped on, they forgot the bet. "We all stared at it and said, 'My god. Do you know what we just saw?'" Human beings, for the first time, now depended more on some other rock's influence than Earth's. Life from Earth had really traveled to a whole different world, as honored by a little, silent bulb.[33]

Other staff monitored blinking lights just to make sure they kept blinking. Engineers babysitting the massive computers running Mission Control recall the telltale sign of computer failure. As they watched the banks of red lights, interspersed with a few greens, they knew a machine had gone belly up when all its lights went on together or all went off together. Since computer freezes were far from rare, NASA maintained three identical IBM computers, constantly updated with mission data, ready to run Mission Control. Engineers ranked computer performance based on how often they failed. During a mission, standings for the most reliable machine could update as frequently as three times per day, and the engineers could swap out which computer ran Mission Control without affecting the mission.[34]

As the Moon loomed ever larger in the windows of spinning Apollo, engineers encountered a new problem. The simulators had left something out, more sociological than technical. "What we didn't do, even later, we never put any [time] delays in," Hughes said. "It's the first time humans had to really deal with the speed of light as a problem." Radio waves, just like visible light, microwaves, and their many spectral relatives, move nearly two hundred thousand miles per second through space. For life on Earth, everything is close enough that we don't usually notice any signal delays. But, once we consider the scale of the solar system, even the sunlight on your skin took a little over eight minutes of transit time to reach Earth. "There was only a second-and-a-half delay [of radio transmissions] to the Moon, which meant three seconds of total delay time. We chose not to simulate that, and we screwed up all over in the real world." The resulting trouble for Apollo? Awkwardness. "We were always talking over each other," Hughes said, "because humans just hate silence." When Houston was done speaking and sending an important message to the spacecraft, they had to wait three seconds before the earliest possible response from astronauts. (Count off three seconds. A person hearing a phone silence that long today would assume the line is dead or that their conversation partner is surfing the web.) For Apollo, as Houston started talking again, they spoke right over the delayed reply from the astronauts. Then the astronauts, after a delay, would start hearing more speech from Houston that had no relation to what the astronauts had most recently spoken.

The solution that evolved for future missions primarily relied on patience with the three-second round-trip delay, but also on a simple verbal tag dating to World War II. Astronauts and Mission Control engineers began saying "over" at the end of their statements, so their far-away partners knew it was safe to start replying. In later Apollo missions, there were extra complications when the lander separated from the other modules. While Houston still suffered a three-second delay for speaking with either craft, each craft could speak to its sibling almost instantaneously. The modules were like two kids jabbering away while sleepy parents, in another room, were slow to understand.

The communications delay becomes a significant psychological problem when considering more far-flung missions. Frank Hughes still tinkers in this area, sometimes giving conference presentations on human considerations for long-term space travel. "As you get further away, the delay increases," he said. "Mars is twenty minutes, but imagine talking to your husband with a twenty-minute delay . . . pretty soon you ask a list of questions and then get a list of answers . . . . It comes down to a really emotionless kind of thing. You say 'I love you,' and throw that in the wind."[35]

Apollo 8 did debut a new and more comfortable sort of space communication. Counterintuitively, it marked the first manned space mission with *days* of uninterrupted radio contact between Earth and the astronauts. The missions in Earth orbit had flitted from one communication station to the next, frequently passing through minutes of radio silence in between. On the way to the Moon, with greater separation, Earth now always had a station facing Apollo. For Moon missions, NASA had fourteen powerful radio stations in play but primarily relied on its three largest radio stations, spacing the globe in 120-degree increments: stations near Goldstone, California; Canberra, Australia; and Fresnedillas de la Oliva, Spain. One of these would always be facing an Apollo spacecraft during its outbound or inbound journeys. The tricky hours in between, however, were a different matter.[36]

For Apollo 8, the most nerve-wracking communications issue was a complete radio black-out as its path slid behind the Moon. At the closest initial approach, the linked modules skimmed just seventy-five miles above a cratered orb measuring over two thousand miles across. If the Moon were a baseball, Apollo eased around it just by the width of a peppercorn. Many Apollo astronauts spoke of the Moon looming larger and larger, and wondering in the backs of their minds if something might be wrong with the calculations. Soon, they could barely see the edges of this enormous, colorless object. Their instincts told them it was too close and that they would surely collide. (My father tells a story of one Apollo astronaut, on a later mission, who couldn't resist this natural animal response and manually rocketed his ship further away from the Moon, just to make sure.)

The mission planners had calculated an initial path that, in the case of an accident or disaster, including a possibly unconscious crew, would sling the spacecraft back toward Earth. However, with all systems functioning and the crew fully awake, Apollo 8 sought a stable lunar orbit. To exit the automatic "free return" path home, the mission had to alter its route on the far side of the Moon, when it was completely out of touch with Earth. The spacecraft burned its service module engine for four long minutes, slowing down and (fingers crossed) entering a new path that repetitively circled the Moon.[37]

The waiting period on Earth, with the tiny spacecraft hidden by the Moon, would become a hallmark for the Apollo missions. In 1968, humans were for the first time completely out of sight and unreachable from all of Earth. While the engine firing lasted just four minutes, the spacecraft itself was out of touch for about an hour.

Those most responsible for computing trajectories—affectionately called "luna-tics" by their colleagues, suffered these blacked-out minutes the most, saying the wait felt endless. Whenever Apollo emerged from the back side of the Moon, the precise timing would tell engineers what they needed to know. If it was exactly on time, then Apollo had found the correct, safe orbit. If it was *late*, they were moving too slowly and risked being yanked to the Moon's surface. And if it was *early*, the craft was moving too fast and risked whipping off to distant regions of the solar system with too little fuel to come home.

On-board, astronauts trained the first human eyes on the far side of the Moon. Luckily, only one of the little windows had fogged over from outgassing sealants. What they saw was remarkable for its savagery. Compared to the near side, the far side looked more weary and pock-marked, but this heavier cratering made sense. With one side locked to face Earth a long time ago, that face—the one we know well—enjoyed modest protection from further asteroid impacts. Meanwhile, its backside remained permanently exposed to whatever the inner solar system dished out. The astronauts would have to wait to share their observations with Mission Control.

In Houston, hundreds of nervous engineers watched the mission timer as seconds oozed past. "I never will forget how quiet that whole room was," said flight director Gerry Griffin of the fifty-plus minutes. "Hardly anyone moved that entire time." It looked more and more certain that the ship wouldn't emerge *early*, but would it be late? Then the radio signal arrived, within a fraction of a second of perfection. Engineers had aimed for an initial lunar orbit of elliptical shape with its low point 60 miles above the Moon and its high point 170 miles above the Moon; Apollo 8 entered an actual orbit of 60.5 miles on the low side and 169.1 miles on the high side, well within a stunning single percent of the calculated goal. In the software laboratory near MIT, engineers broke into ecstatic cheers. Apollo's on-board computer had handled the maneuver, with little corrective input from Earth.[38]

A reporter recalled the moment from Houston.[vi] As he monitored the Mission Control center, "Suddenly the familiar map of the Earth vanished from the big plastic screen." This had been the norm for the last decade, with all space missions orbiting our home planet. But now, in its place, and with no fanfare, a practical map of the Moon dropped down. "The effect was overwhelming." The reporter, a former war correspondent, said it was one of the most memorable "heart-stoppers" of his career.[39]

After a couple of extra adjustments, the ship entered a more circular orbit and started collecting information on the Moon's uneven gravitational pull. "Your orbit almost wiggled around the moon rather than a circular orbit," engineer Hal Beck said.[40] The mission completed ten orbits of the Moon in about twenty hours.

[vi] Robert Sherrod.

Astronauts noted how much more slowly they moved above the Moon than when orbiting Earth, and this followed from basic physics and weaker gravity.

On Christmas Eve, the Apollo mission sent a formal broadcast to Earth, where an estimated one *fourth* of humanity huddled near a radio or television to hear a message from the Moon. The mission had already sent back the incredible "earthrise" photograph of our tiny blue planet, peeking above the lunar horizon and suggesting a reset perspective for our species. It showed a fragile orb in a largely inhospitable dark cosmos. No conflicts or boundaries or protests or gunshots could be seen from the Moon. No person looking at this photograph could deny that all people (except for three astronauts) were in it together.

The broadcast gave most of NASA a surprise. Before the launch, lead astronaut Frank Borman had wondered what he could possibly say, from the Moon, that would rise to the singular moment. He had appealed to some friends for help. One of them recalls staying up, long into the night, skimming books, trying to figure out something to recommend to the astronauts.[vii] He eventually started leafing through the Bible's New Testament. His wife walked by, checking on him, and said he might want to try the Old Testament. And so, on December 24, 1968, Earth took a break from a confusing and turbulent year to hear astronauts reading the Christian origin story of the planet Earth. Each astronaut read a few verses, with William Anders starting with "In the beginning," and finally Borman concluding with God naming "the dry land Earth."

Another of Borman's earthbound friends was traveling at the time. "That night in the airport lounge, as the astronauts read from Genesis, there wasn't a single word spoken. Not one. I'm not sure that would be the case today." He lamented that the moment's magic could not be bottled and treasured. "Things sort of lose their currency, after about twenty or thirty years, because the up-and-coming generation has no capacity to understand it." And despite the peaceful phrasings and the calm voice crackling from a quarter-million miles away, many on Earth still felt a ghostly tension. "When they did that reading," the friend recalled,[viii] "they still had to make the burn to get back home to Earth. We didn't know for sure if they'd make it back or not."[41]

For weeks ahead of time, media coverage had included a morbid fascination with the next crucial step. Apollo coming home meant rocketing away from the Moon's pull. To do this, astronauts would restart the single engine on the service module.

[vii] Joseph Laitin, who served the press communications efforts of five presidential administrations.

[viii] Simon Bourgin, who served as science advisor for the U.S. Information Agency during Apollo.

The engine had no back up, and if it failed, the crew would be marooned circling the Moon, slowly running out of air. But most of the systems *within* the engine had back-up systems. As one engineer who worked on it said, "There were a heck of a lot of ways to start that engine." And yet, now that the engine was so far away, with three lives attached, "It was darn scary."[42] The engine restarted without so much as clearing its throat, and Apollo 8 headed home.

Of all space's dangers, from the extreme temperatures to the ever-present pull of the voracious vacuum, Apollo engineers also had to plan for solar flares: the sun's infrequent ejections of radiation (and gobs of high-energy ions) that could give astronauts a hazardous zap. Solar flares could erupt with little warning, and the fastest components would hit Apollo in eight minutes, but engineers had plotted a strategy. The plan had two parts. If possible, stay in orbit around the Moon, since its bulk would protect the astronauts for half of each orbit. Then, turn the service module toward the sun, providing the astronauts additional shielding from the nasty radiation and particle streams. Apollo 8 encountered no flares, and nearly returned home without incident.

The lone hiccup involved Apollo's on-board computer. As with many computer errors, this one followed a user's mistaken keystrokes. For the return to Earth, the crew worried about perfecting their precise angle of attack for the atmosphere. They were moving faster than any previously re-entering ship, and the command module was also larger and heavier than the earlier capsules. But one astronaut mistyped a command and erased all basic navigational data. Instead of making a minor correction to their trajectory, he had ordered the computer to prepare for launch, as if it sat on the pad in Florida. So, following orders, the computer lost its basic notion of up from down, and the crew scrambled to the windows, trying to locate their guide stars. A cloud of urine crystals, flying alongside like so many little geese, made the starscape illegible.

Luckily, engineers in Mission Control had most of the two-plus-day trip home to diagnose the problem and talk the astronauts through re-aligning the computer. In building the system, some engineers had originally wanted to have more warnings for possible "operator errors," such as today's "Are you sure you want to delete this?" message, but the Neolithic machine was already pushed to the limit of its small memory and modest processing power. (A later version incorporated a kind of "undo" option for the astronauts.)[43]

As the mission approached Earth, the onboard computer correctly aligned the capsule and they entered the atmosphere at the appropriate three-degree angle (where zero degrees would be skimming across the atmosphere's ceiling on a tangent line). Like previous capsules, the Apollo command module obeyed Faget's original idea, facing its wide bottom forward, with the astronauts on their backs and facing the heavens. As this capsule slowed from its record speed, the braking acceleration pulled the astronauts back into their seats with seven times the force of gravity. Each of them suddenly gained the weight of an adult sea lion on their human bones. One astronaut became alarmed when he saw fist-sized glowing "blobs" flying past the windows. Was the heat shield losing that much material? But each tiny bit of heat shield that flew off (as the ablative material was designed to do) now became a burst of glowing plasma at their incredible speed—beautiful pyrotechnics but no cause for alarm this time.

The violence of re-entry and the accompanying plasma meant the command module went into communications blackout for the first time since it emerged from behind the Moon. Engineers sweated the last three dangerous minutes of a long mission. Underneath all their excitement in preparing and then monitoring Apollo 8, Frank Hughes recalls a special human gravity surrounding a mission full of unknown territory, including unprecedented distances and speeds. He'd gotten to know some of the astronauts in the simulators and in star chart training. Before that December launch, astronaut Bill Anders had asked Hughes if he would make sure his wife got home safely to their home near Clear Lake, come what may. Frank had promised he would. "It was so solemn," Hughes said later. "It was like you're making these last will and testament kind of moves."

After six days of hurtling through space, the Apollo 8 capsule finally stopped, coming to rest in the Pacific. Astronauts immediately noted pooling water at their feet and worried that seawater was seeping into the capsule. But the intense atmospheric braking had simply thrown every tiny water droplet, from a week's worth of breath and spills, to form a big puddle at the capsule's floor. If anything, their approach had been too perfect. The capsule hit the sea within just two miles of an aircraft carrier, arguably too close for comfort. They would allow for greater separation from now on, making sure to avoid a catastrophic on-deck landing.[44]

With a surprising number of Apollo engineers, the last months of 1968 still resonate deeply, with a sustained sense of awe. Many rank the period, without hesitation, as the program's peak. They'd flown three humans all the way to the Moon and returned them not just alive but with no obvious injuries. In retrospect, American and Soviet engineers alike mark this mission as the end of the space race.

Just eight years earlier, a little farm-to-market road ran from Webster, Texas, toward Clear Lake, bordered by nothing but cows and barbed wire. As of December 27, 1968, the road exploded in a no-holds-barred party, with revelers dancing across the pavement. In a nearby hotel, drunken engineers threw furniture into the swimming pool. After this trip to the Moon, with days of television coverage showing astronauts and Mission Control, Houston became the leading character of NASA's public relations. And despite the enormous success of the Saturn V rocket—in only its third launch, it had pushed Apollo to the Moon—Huntsville receded in terms of prestige. "Compared to astronauts," von Braun said during a Houston visit, "our Saturn has about as much sex appeal as Lady Godiva's horse!"[45] Just ten years prior, von Braun and his strong chin were cover-worthy for *Time*; for 1968, the magazine pivoted from their planned cover (reportedly president-elect Richard Nixon) and named the three Apollo 8 astronauts together as its "man of the year." The astronauts received bushels of fan mail, but one struck a famously pithy note: a telegram simply thanked them for "saving" an otherwise awful year for the planet.[46]

After the safe return of Apollo 8, one NASA administrator told the press that the mission was "the triumph of the squares—the guys with computers and slide rules who read the Bible on Christmas Eve." The statement bristles with pride to the point of being prickly or even cutting. *Time* magazine explained the sentiment when the editors opined that NASA engineers "epitomize the solid, perhaps old-fashioned American virtues." In a late-year essay titled "Of Revolution and the Moon," the magazine declared, "This is what Westernized man can do. . . . [H]e will not drop out and turn off [*sic*]; he will not seek stability and inner peace in the quest for nirvana. . . . [H]e knows how to reach the Moon."[47] With the counterculture movement in full flower, some of NASA's engineers, having started this intense work in 1958, could hardly recognize the country when Apollo 8 came shooting back to Earth just ten years later.

12

1969—ALARMS AND LIGHTNING

He was sitting with almost a hundred and fifty other villagers in the assembly room of the new parish hall, watching two young Americans walk on the moon for the first time. There was a tense silence in the room for almost the whole of the broadcast, yet . . . everyone started cheering, and for a few moments at least it was as if some kind of burden fell from the farmers' heavy shoulders. Afterwards there was free beer for the adults and juice and doughnuts for the children, and a member of the parish council gave a short speech about the tremendous endeavours that made such marvels possible and would probably drive humanity on goodness knows whither.

—From the novel *A Whole Life*, by Robert Seethaler, set in the Austrian Alps

I headed up the river, past dusty villages so unused to the new American presence that young women bathing in the river saw no need to cover their breasts as they waved at the boats. Little children ran to the water's edge repeatedly shouting something that sounded like Ah-Pa-Lo, which baffled me until I realized the word was Apollo. The moon landing was still our greatest ambassador there. They didn't know to distrust us yet.

—U.S. Naval Officer Armistead Maupin's account of an expedition into Cambodia, from his memoir, *Logical Family*

At the start of a most crucial year for Apollo, two giants of the program focused, to a surprising extent, on anything but the Moon. Max Faget's son now recalls that his father brought home a new model at about this time. "He cut all the little balsawood pieces, and he made an early mock up," Guy Faget says. "He and I went out in the back yard and played catch with it and watched it land in the grass." Father Max had been talking about this idea for years now. "It's going to be lifted into orbit by a rocket," Max told his family. "And it's going to glide back." Henry Pohl remembers Faget bringing versions of his new spaceship to staff meetings and gliding them across conference rooms. Dating to the 1950s, Faget and others had wondered about a winged craft: a spaceship sophisticated enough to maneuver in orbit and then survive a fiery re-entry, returning to Earth as an airplane. The idea of a "shuttle" running to orbit and back could, NASA believed, greatly cut the cost of spaceflight, by reusing more components.[1] (See Figure 12.1.)

Wernher von Braun, as attuned as anyone to the increasingly pessimistic budget forecast, supported the notion of a shuttle program as a way to revolutionize the cost

FIGURE 12.1 Max Faget, in 1969, with an early model of a reusable, gliding spacecraft—a shuttle.
(NASA photograph.)

structure of space flight. He advocated not just for some sort of winged shuttle but
also for a winged, reusable rocket to be developed in Huntsville. In this picture, *everything* would be reused: land it, clean it, refuel it, and launch it again. NASA, bright-eyed with recent Apollo success, claimed they could lower the costs of lofting things
to orbit from Apollo's five hundred dollars per pound to just fifty dollars per pound
(an absurd dream, in retrospect).

   To keep Huntsville in the game, von Braun maneuvered his center for new space
work in the post-Apollo era. Aside from lifting a shuttle craft, he'd also embraced an
emerging project called "Voyager," that in its early incarnation aimed to send sophisti-cated unmanned landers to various planets in the solar system. Marshall was to lead the
way and had requested a significant new budget line in 1967, but Congress had nixed it.

In 1969, NASA headquarters kept its chin up and expressed optimism against all federal headwinds. As von Braun wrote in a letter to his father, they hoped this new president, Nixon, would want a robust, achievement-focused space program; headquarters asked von Braun to spearhead their D.C. presentations advocating for Mars. Vice president Agnew, for one, was a vocal proponent. But public polling and congressional sentiments should have discouraged NASA from the red planet. Though von Braun's heart shouted "Mars or bust," he was also showing a realistic side in this era. "I, for one, have always felt that it would be a good idea to read the signs of the times and respond to what the country really wants," he told the political scientist John Logsdon, "rather than trying to cram a bill of goodies down somebody's throat." He knew Mars was a tough sell. Instead, he started speaking passionately for projects like Earth observation and new communication satellites that could show immediate benefits to humankind.

In fact, von Braun had confided to family members by 1969 that he had tired of directing the Marshall Spaceflight Center. It is less likely that the long hours of 1957–1969 had dented his nearly inhuman energies, and more likely that he recognized America's reduced appetite for space. He'd already watched his Huntsville stronghold change, as retirements and transfers depleted its hungry optimism. He started talking discreetly to NASA leadership about new possibilities within the agency, perhaps turning loose his powers of persuasion—his scientific charm—directly on Washington, D.C. After all, there had never been—and really, there never would be again, with all due respect to Carl Sagan—a better salesman of space.[2]

Though the Houston-versus-Huntsville rivalry was alive and well in the era of budget reductions, von Braun and Faget, the two dreamers, had learned to appreciate one another. Recognition of mutual interests grew during their 1967 trip to Antarctica (a naïvely optimistic scouting trip to envision requirements for lunar bases and Mars missions). Max Faget's son, Guy, recalls the family hosting von Braun one evening circa 1969, as the former Allied submarine officer and the Axis weapons designer celebrated their Apollo success on the Dickinson, Texas, bayou. "Our Mom was very taken with him," Guy says of their tall, dapper guest. "Very charming guy . . . loud, pleasant laugh." Max asked Guy to show von Braun his model rocket collection. After paying patient, close attention to the boy's presentation, von Braun praised Guy's attention to detail.[3]

While these visionaries looked forward, most of NASA toiled to realize a previous and now pressing dream. As of January, 1969, engineers still had to pull off the unprecedented within twelve short months: landing humans on a foreign body, where any substantial mishap could publicly extinguish three lives.

Preparing for Apollo 9, technicians and astronauts crawled into the latest lander on Earth, but despite tiptoeing about the aluminum eggshell of a ship, they broke components at every turn. NASA had one gaping question: Would a lander hold its various cracks and fragile wires together long enough to land on the Moon? The program had yet to put a habitable version into space.

After various bouts of binging and purging, the lander had stabilized at about 3,300 pounds, the weight of a midsize sedan in today's world. But the lander distributed those pounds like so much cotton candy. It stood twenty-three feet high and its legs spanned an area thirty-one feet across. The bug's crew cabin had many times the volume of a sedan's interior. Its rows of controls, switches, and indicators would overwhelm even the busiest of today's dashboards. And unlike a car, the lander had to protect two humans from the hazards of space, all while boasting an extra engine fit to lift itself from the surface of a foreign body.[4]

In January, NASA busily prepped Apollo 9. In this mission, engineers sought to place a lander into Earth orbit, along with the command and service modules, and have them practice their intricate dance near Earth: a dress rehearsal before taking the show to the Moon. Owing to basic issues of geometry, the bulbous lander could not launch in its final form, with legs deployed, nestled atop a Saturn V. During launch, the pointy-headed command module sat on top, with its service module right behind it. Underneath those sat the lander, with its legs folded like a dead, dry spider. Assuming everything went well with the launch, the astronauts would fly the command and service modules, as a combined cone-plus-can unit, away from the remnant third stage of the Saturn V; the folded lander would wait atop the third stage. Spinning carefully around to then face these remnants, the astronauts would move the pointy nose of the command module forward and dock with the lander. The two very different craft—one simple, conical, and sleek, but the other awkward and lumpy—were then connected head to head. And this unlikely chain, starting with the can and engine of the service module and ending with the legs of the lander, would comprise the entire Moon-bound Apollo spaceship. For the ninth Apollo mission, NASA wanted to build this chain in Earth orbit. Then, once astronauts occupied both the lander and the command module, they would separate, maneuver, and then dock once more.

With some trepidation, a team from Houston arrived at the Cape and meticulously checked through the latest lander. They found nearly four hundred splices inside, where Grumman workers had already mended broken wires. Biting its collective lip, NASA decided this would have to do. They really needed to give the bug a chance to fly.

The more significant worry involved docking between the top of the command module and the top of the lander. The three-dimensional trials in Houston had shown the connectors to be finicky. Even if the probe of the command module found the drogue of the lander, the latches sometimes failed to secure their union. In other cases, a partial connection led one module to essentially throw the other over its shoulder, as if in a judo match. In space, this unwelcome move could be a death blow.[5]

Meanwhile, the USSR trumpeted a successful January docking mission, as the Soyuz 4 and Soyuz 5 spacecraft connected in orbit. Two cosmonauts exited one ship, and spacewalked to the other. In an eerie moment, one cosmonaut lost his grip on a camera. Instead of documenting their historic transfer, the camera drifted away, in a slow tumble, into the star-speckled void.

After separating the craft, three of the cosmonauts safely returned to Earth aboard the Soyuz 4, while the last cosmonaut waited to bring down Soyuz 5. His descent to Earth, held for many years as a state secret, puts any dramatic episode of Star Trek to shame.[i]

As with NASA's program, the Soviets now had modular spacecraft, and the descent capsule separated from a supply- and instrument-rich segment that stayed behind in orbit. However, on Soyuz 5, they didn't cleanly split. (It could well have been a problem like the one Pohl and his young employee found with the incomplete line of explosives meant to cleave Apollo modules, or one like the finicky docking latches.) The cosmonaut had no choice but to set his jaw and attempt re-entry dragging a non-aerodynamic, unwanted hitchhiker. As soon as he began slipping into the main atmosphere, this extra appendage tripped on the rushing air and the entire ship began to tumble end over end. Mercifully, the ship stabilized, but the cosmonaut realized to his horror: it had flipped backward. His heat shield now faced the stars. Instead of the dramatic slowdown pushing the cosmonaut into his seat, everything was reversed—physics now yanked him away from the seat and tried to extrude him through his restraining straps toward the nose of the capsule.

Unprotected by a heat shield, the top of the ship began to melt, and smoke snaked about the capsule's interior. The cosmonaut heard tanks of unused propellant exploding in loud bursts outside. Soon, the braking craft pulled him against his straps with more than nine times the force of gravity. And he could feel the intense heat growing all around him. He might dissipate as a bright meteor over the Motherland.

In a stroke of luck, the chaotic rush of air spun his craft quickly about face, so that his heat shield could finally do its job. But now, pushed back into his seat at last, he realized that the parachute casings near the smoldering nose had been welded shut. Without the normal slowing from the chutes, the capsule slammed into the ground.

---

[i] We now know the less positive stories of the Soviet space program from archival material freed during and after the *glasnost* period of the 1980s.

Miraculously, he survived. The jarring impact broke many bones, but he was alive. In the sudden quiet, his steaming ship hissed against the frigid Russian soil.[6]

While not announcing this terrifying re-entry, the Soviets did plan a massive celebratory parade for the four cosmonauts, claiming that the docked spacecraft had formed "the world's first space station." They were hungry for PR success after the Americans' Christmas lunar trip. But the turbulent 1960s could reach any city, including Moscow. During the parade, a soldier in dark glasses stepped from the crowd and drew two pistols. Angry at being conscripted to the Russian army, he sought the life of General Secretary Leonid Brezhnev, but he began firing at the wrong car. Before police could tackle the gunman, he killed the driver but missed the cosmonauts, who dropped to the floor under a spray of broken glass.[7] They had to wonder if life on Earth was becoming more dangerous than life in space.

An American at the time could have understandably considered the same question. Against the backdrop of a growing antiwar movement and both political parties preaching "law and order," America continued to fracture its view of technology. In 1969, Housing and Urban Development, an agency even younger than NASA, lured away one of the space agency's lead administrators, Harold Finger. Becoming HUD's "assistant secretary for urban technology and research," Finger led the new "Operation Breakthrough," aiming to use technology and engineering approaches to somehow cure urban violence through the structures of cities themselves. The effort (ultimately futile) envisioned a sort of pre-fabricated, calming set of Habitrails for a new urban America.[8]

While some wanted technology to fix every type of social trouble, others continued turning away. In 1969, aviation pioneer Charles Lindbergh published what would become a very popular letter, renouncing technology entirely. The sixty-seven-year-old said technology had lost its way in America and that people must embrace a more natural, intuitive path to living on Earth. After serving as a boyhood hero to so many future NASA engineers, he now shook his head at the approaching Moon landings, telling readers that "in instinct rather than in intellect is manifest the cosmic plan of life." In essence, set down your protractors, America, and go with your gut.[9]

In early March, NASA launched the ninth Apollo mission on top of the fourth Saturn V rocket. This time, ascent to orbit went smoothly, give or take a bit of pogo in the second stage.

The command and service modules glided away from the Saturn's top stage, and now Apollo just needed to spin about and pull the lander from atop the Saturn's third stage. But the command module was responding oddly and acting sluggish. Engineers on the ground diagnosed the problem—as he floated freely, an astronaut had accidentally flipped a switch and deactivated a set of thrusters. With a sigh, the engineers made a note to add more switch guards.

Docking with the lander went surprisingly well for a first attempt and now all three modules, for the first time, flew as one Apollo craft in space. As two astronauts crawled into the lander from the command module, one astronaut became incredibly space sick, reacting to his body's new absolute freedom. Engineers had to revamp their mission schedule, particularly for his spacewalk. Sudden vomit inside a helmet (without helpful gravity to pull it footward) could clog an astronaut's airways, leaving his gloved hands pressed uselessly against the helmet's glass.

Aside from one slightly green passenger, the mission suffered very few technical worries. Over the next ten days, the lander performed well, and NASA rated the mission a success. Astronauts complained that the lander was a noisy little thing inside, but it survived jostling, docking, and maneuvering in space. It did everything but land on something—that would have to wait, but hopefully not long. The decade had barely eight months remaining.[10]

The "mobile launch" concept entered full bloom at the Cape. Before the ninth mission had crawled toward the launch pad, technicians had already stacked the next Saturn V rocket in the vaulted assembly building. By the time of the next mission, in mid-May, four manned Apollo launches had spanned just seven months.

But sometimes a head start could create its own complications, leaving too much time to tinker. Approaching its launch date, the tenth Apollo mission finally arrived on the pad, but a technician decided to replace a suspect valve. Cutting the rocket's power line to make this change triggered part of the fuel system to open its own valves. Thousands of gallons of kerosene immediately gushed from the giant first stage and spread over the launch pad. The tank itself, roughly the size of Henry Pohl's childhood cabin, started to buckle inward without the pressure of its contents. Happily, no stray sparks found the wayward fuel, and engineers avoided a pad conflagration. Refilling the thin metallic balloon of a fuel tank even kicked out most of its new dents. With a nervous shrug, engineers decided to plow onward, restarting the countdown.

Launching such a major mission every two months made for a sort of steady mayhem in Houston. Engineers separated into teams for planning, preparing, and simulating the staggered missions, each with different aims, different computer programs, and even different equipment. In interviews, they speak of sometimes wandering from real missions in Mission Control to simulated practice missions in a nearby parallel facility, and becoming confused as to which was which. Engineers working a fake, preparatory mission were just as serious and focused as those running the real thing.

Planners created recipe lists of commands for astronauts at every stage of a mission, including sequences for various emergencies that would hopefully never come to pass. Engineers referred to the extreme game planning as "dispersion analysis"—the many forking paths of the future—and some worked for years on alternate scripts that remained forever shelved, happily never needed. But mistakes were inevitable in these growing stacks of plans and procedures, no matter the double- and triple-checking. And a simple mistype on one such list snuck aboard the tenth Apollo mission.

Departing in May, Apollo 10 would return our species to lunar orbit. This time, NASA aimed to practice everything *but* the landing, echoing Faget's original, informal answer to how one approaches the Moon: cautious step by cautious step, adding one complication at a time. The launch was a rough one, as the second stage began with an intense kick—one astronaut said it felt like "a train wreck." A few minutes later, pogo oscillations began with such intensity that the astronauts considered aborting the mission. Eyes shook so fiercely that they could not read shaking instruments. One astronaut forced his quaking hand toward a manual abort handle, but decided against it. They survived the pogo, entered a serene orbit of Earth, and soon they refired the third stage engine to push for the Moon.

Only at this point, at their maximum speed, did the Apollo modules begin their elaborate dance, where the command module spun about and gently plucked the lander away from the third stage. Soon, all three linked modules headed for the Moon, after a little thruster kick pushed the depleted Saturn stage away (see Figure 12.2).

It may sound absurd to think of the modules executing delicate maneuvers, including docking and extraction, while moving at about twenty-five thousand miles per hour, but this perfectly underscores what Einstein (and Galileo before him) saw as the lack of any real meaning for an object's speed. Any movement boils down to: with respect to *what?* In this case, because the command and service module, the lander, and the third stage were all moving together, they had no speed versus one another. And space has no wind to blow one's hair back. While the Apollo speeds sound ridiculous, all major astronomical bodies have large speeds relative to one another, compared to our tame, everyday speeds on Earth. Earth moves at about seventy thousand miles per hour as it orbits the sun. And the sun, in turn, moves at about five hundred thousand miles per hour around the center of the Milky Way galaxy. These *relative* speeds have meaning, but since there is no absolute fixed point—no central rock, or star, or black hole in the universe—there is likewise no way to assign absolute speed. In this sense, Apollo setting a "speed record" for humanity should be qualified: the greatest speed of human beings . . . with respect to Earth's surface.

After docking, the first astronaut entering the lander pulled up short. A flurry of little white bits floated in to greet him. They lit on his spacesuit, his hair, and his face. The lander's walls had shed insulation at some point, perhaps during the

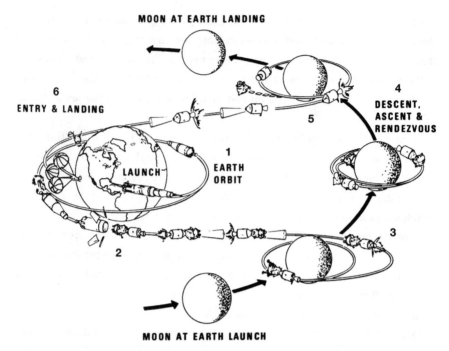

FIGURE 12.2 A step-by-step Apollo Moon mission. The Saturn V, topped by the three Apollo modules, (1) launched into Earth orbit. After the third stage re-fired and pushed the mission toward the Moon, the command and service modules spun about, docked with the lunar module, and (2) pulled it away. (Subsequent "barbecue" mode not depicted.) Once arriving at the Moon, (3) Apollo rotated until its service module engine could slow it down to enter lunar orbit. The lunar module would then depart, (4) head for a landing, and later return to rendezvous and dock in lunar orbit. With the astronauts all reunited in the command module, (5) Apollo jettisoned its lunar module, and the service module's engine then propelled the mission toward Earth. The cones in the front of the ships give a sense of the engineers' notion of "dispersion," a range of possibilities. Coming back to Earth, (6) the mission would jettison the service module, bypass Earth orbit, and re-enter the atmosphere. (The labeling has been modified from a 1969 NASA illustration. Almost nothing here is to scale.)

cocktail-shaker launch.[11] What else might have come loose? Luckily, engineers had wired the newer landers with slightly thicker wire. It carried extra pounds but had comforted engineers with many fewer problems.

This trip to the Moon showcased engineers' calculations like never before. The two hundred forty thousand miles went by like a perfect computer simulation. Given a flawless trajectory, engineers skipped most of the scheduled checkpoints for course corrections. Once our nearest neighbor loomed large once more, three humans slipped away from any view of the Earth and around the back side of the Moon, where a perfect burn of the service module's engine inserted the mission into lunar orbit. With two astronauts aboard, the lander departed its sibling modules and descended to just nine miles above the lunar surface. Skimming over the Moon this close allowed NASA to get the best data yet on the uneven gravity there. If the two astronauts were tempted to defy orders and set down on the Moon, self-preservation stopped them. Figuring it was best not to tempt anyone, engineers hadn't loaded enough fuel for a landing.

Flying over dramatic, colorless landscapes, the astronauts went through their lists of orders for flipping switches and entering commands. They had done this thousands of times in practice missions and simulations, with no reason to doubt what was typed on a page of instructions. But suddenly a crew member radioed a very unpracticed and un-NASA-like "son of a bitch!" to Earth. The astronaut, Eugene Cernan, didn't even append "over." The lander, now spinning out of control, seemed to have lost its computerized mind.

As the astronauts in the low-flying lander had prepared another in a long line of tests, one of the recipe lists instructed them to set a single switch—one of hundreds—incorrectly. Following the mistaken setting, the lander had commenced an automated search for its partner modules. But they were out of sight, in a different orbit. It was as if a stranger had leaned down to a grocery-store-wandering toddler and sternly asked, "Do you know where your parents are?" A violent pitching rotation resulted, with alarm lights flashing. The lander warned the crew that it was about to lose all notion of its own orientation—it wouldn't know which way was "up" from the Moon—and the sickening spin might become permanent.[ii] Astronauts quickly took over manual control and righted the crazy ship, but not before the expletive made its one-and-a-half-second sprint to Earth's waiting ears, alarming Mission Control's assembled engineers (and probably a few Russian eavesdroppers as well).

After successfully redocking with the command module, the three astronauts, together again, completed thirty-one orbits of the Moon and refired the service module engine for their return trip. The entire eight-day mission, despite extra spins and stress, solidified confidence in the Apollo game plan and brought home the best description yet for the Moon's spastic gravity.[12]

Rolling into July, with every preparatory step complete and every module working well, NASA readied a Moon landing. Just eight years and two months had passed since President Kennedy issued the challenge. The milestones, the solutions, the hours, the sum toll on the engineers and their families didn't seem to fit within that time. But it looked like all the work might pay off with a hard-to-fathom moment. Henry Pohl's son, then about five, recalls big gatherings of the engineers and their families, as NASA welcomed them to the center and screened films summarizing each Apollo mission along the way.[13]

[ii] They risked a condition known as "gimbal lock," where two of a craft's three gyroscopes become twinned, essentially removing one full dimension of awareness.

The media prepared for a grand show. As the Apollo 11 launch approached, 3,500 reporters from around the world descended on the Cape, and hundreds more gathered near Mission Control in Houston. At a July 10 press conference, a NASA panel fielded all manner of question about astronauts, the spacecraft, and the Moon itself. But the once-central Wernher von Braun received exactly zero questions; he sat politely listening. Given the chance later to make some remarks, he responded to all the great excitement and what was, to some, a celebratory air. Maybe the lack of attention made him a bit cranky. But he knew they'd been lucky in space so far. And surprises surely lurked in attempting a landing. A sobering number of unmanned probes had accidentally smashed themselves on the Moon like so many dropped pieces of china. "Something may happen and the public should be prepared for the shock," he said. And even if the astronauts lived, he wanted to temper excitement for the maximally safe and relatively featureless landing site and what could be learned there. "The *maria* on the Moon," he said, referring to the flatter, darker areas, "are about as much value to science as a gravel pit."

When a clutch of reporters gathered around him on July 14, he let himself return to his more quotable rocket-sage self. Asked how to place a Moon landing in historical context, he ranked it "about with the importance of aquatic life first crawling on land."[14]

CBS News prepared a television extravaganza of wholly new proportions. They built a massive studio especially for the Moon landing, with 142 cameras, a special interview area, a new desk for mission updates, and a number of spacecraft models. Space fan Walter Cronkite would run the show, planning for an unprecedented day-and-a-half of *continuous* coverage. But the ringmaster behind the scenes, producer Robert Wussler, pioneered a new sort of broadcast. How would CBS fill all those minutes? For one, they would roll out "experts" as talking heads. Wussler hoped academics, astronauts, literary figures, and NASA officials could jabber away between infrequent bursts of updates. CBS lined up hundreds of candidates ahead of time. A decade later, Wussler would co-found the Cable News Network, utilizing the same blueprint, one that would eventually fill our lives with armies of arguing faces.

As additional filler and entertainment for audiences, the network enlisted the services of a small animation company, Reel III, to create life-like visions of the space mission. Their short movies, incredible in retrospect, offered vistas that could never be captured by a real television camera (e.g., the lander detaching from the command module in lunar orbit, or a perfect perspective view of the lander approaching the Moon). A technical illustrator named Ralph McQuarrie spearheaded these animations. After modeling the gritty reality of the function-over-form Apollo missions, he would go on to advise a young director on a space epic called *Star Wars*, with its own lumpy and gritty spaceships.[15]

Engineers went through countless checklists, and technicians monitored hundreds of dials for fuel pressures, temperatures, and boil-off rates. Still others worried about the sun. It had suffered an unusually restless month, having spit forth a couple of solar flares already. Whether in barbeque mode or not, an Apollo craft in transit could get roasted by such an event.[16]

Engineers kept running simulations into the last weeks before takeoff, trying to find each conceivable thing that could go wrong and combing computer code for whatever remnant bugs might be lurking there. Day to day, their confidence improved that this mission really would work. Mission Control managers told a small group of engineers to make sure they understood every possible Apollo computer alarm and what it meant. The engineers devised robust "cheat sheets" telling them what to do for each error message, even the most unlikely ones. "As I look back on that, I'm horrified," said engineer and Apollo computer guru Jack Garman, "because we'd have them written on the back of envelopes. . . . We would stick these things under the plastic on the console."[17]

Some of the errors were head-scratchers. In one case, just days before launch, Apollo's computer appeared to throw up its hands and give up. Engineers failed to diagnose the problem or what had triggered it. They simply had to abort that practice run—what else could they do? They hoped that obscure error wouldn't show up again during the real mission. Consulting with the programmers at MIT's Instrumentation Lab, the Mission Control crew learned that some of the error alarms could be ignored, while others were dire. They annotated their lists accordingly.

Engineer Tom Moser, sitting at his desk in Houston, received a strange visit from his boss. "I'm going to give you an assignment, but you can't talk to anybody about it," the supervisor said. "Congress has said we're putting a United States flag on the Moon." This was against the spirit of a recent treaty, but Uncle Sam was going for it.[iii] Now, this one engineer needed to design a storable flag that could easily deploy and fly proudly, in a vacuum. Moser would also need to find a place for the flag. The boss emphasized there was no spare room inside the lander. Stored somewhere outside, the flag had to be easy for astronauts to find but not close enough to the engine to get toasted. Moser came up with a telescoping flag stored in a long, thin case, almost like a pool stick. A top bar would hold up the nylon stars and stripes, despite the absence of breeze. After a lot of nervous consideration, he decided to attach this case to the lander's descent ladder. He determined it shouldn't affect the ladder's strength or

---

iii The Outer Space Treaty of 1967, originally signed by the United States, the United Kingdom, and the Soviet Union, set a framework for international space law. It barred participants from installing weapons of mass destruction in space, establishing military bases in orbit or on the Moon, or claiming sovereignty over another celestial object.

performance, but given the secrecy, and the short deadline, he couldn't have anyone double-check his work.[18]

As the launch approached, America wasn't exclusively flag-waving. Various protests ramped up. An engineer recalled leaving a long Houston shift and seeing bona fide hippies protesting nearby. He'd heard about protests in the news but had never actually seen one before. This one appeared subdued in the thick summer heat. And at the Cape, civil rights leader Reverend Ralph Abernathy declaimed technological extravagance by leading a mule train, a symbol of long oppression and poverty, to the launch complex. A NASA administrator went to meet with Abernathy, suggesting earnestly but not literally, "I want you to hitch your train to our rocket." We're all in this together, he wanted to say. The reverend wasn't persuaded, but he politely accepted tickets to watch the launch from the viewing stand.[19]

On July 16, an incredibly humid day even by Florida standards, the Saturn V rocket did not disappoint the crowds, and better yet, every stage gave the precious cargo a relatively smooth, pogo-free ride to orbit. The big hydrogen engine on the third stage relit on request and sent this eleventh Apollo mission on its way to the Moon. With all modules flying away from Earth, the astronauts maneuvered to dock with the bug and pull it from its shroud. The only slight cause for worry reminded some of the tragic capsule fire: an astronaut crawling into the lander detected a faint, sour odor, as if something might be smoldering.

The two-day trip to the Moon went smoothly, with just one course adjustment en route. After Apollo eased into lunar orbit, the astronauts, their spacecraft, all of NASA, CBS, and Walter Cronkite were ready, and the lander detached from its siblings. With its crew of two, it finally began its real work, heading for the lunar surface. Its arc was not straightforward. Astronauts could not stand at their windows, watching the approach of craters and chasms. For most of its descent, the lander glided at a shallow angle with the astronauts facing up—imagine sledding down a gentle slope on your back. The crew faced the stars, placing their faith in their instruments and their onboard computer.

Many busy systems on the lander came to chattering life, making sure it knew where it was, exactly, and where the Moon was, exactly. Its single descent engine, wholly separate from the one that would later propel it back to orbit, was a flexible machine, angling its thrust much like a skunk does its precise sprayer. This wasn't just for fine control of the lander's descent. It was a lumpy craft, and it contained not just two fidgety humans but also steadily decreasing fuel levels. In short, it was not what physicists call a "uniform body." Hence, the lander in action was prone to unwelcome spinning, if an engine burst wasn't directed through the lander's exact center of mass. As the lander's distribution of weight constantly shifted during descent, the lander's engine nozzle, guided by the on-board computer, would thrust in such a way as to avoid spins.

Meanwhile, two sets of radar nervously chirped. The landing radar sent four beams downward toward the Moon. One beam regularly measured the distance to the Moon, but each of the other three reported the lander's speed (relative to the surface) in three directions: forward-backward, left-right, and up-down. These three employed the same "Doppler effect" used by highway patrol officers and meteorologists to measure speeds of cars and winds. A separate radar system aimed in the other direction, away from the Moon, ideally just to track the command module when they later docked.[20] The lander's guidance computer had a lot to juggle. (In Huntsville's U.S. Space and Rocket Center, you can try a video game version of piloting a Moon landing—I was zero for four on the easiest setting before I noticed that a line of unhappy kids waited their turns behind me.)

As von Braun rightly predicted, this first manned attempt was bound to have surprises. Astronauts had trouble keeping the lander's antenna pointed at Earth during their careful descent. The Moon-to-Earth signal came and went, but this was merely an annoyance compared to the real problems looming.

"When one of these alarms came up, it would ring what was called the master caution and warning system," Jack Garman said, in describing what happened next. "Like having a fire alarm go off in a closet."

At about thirty-five thousand feet—the altitude of a commercial airliner—the astronauts reported bad news to Houston. "They get this four-digit code for what the alarm is, 1201, 1202 were the two alarms," said Garman. (Upon his eventual retirement, his friends gave him a T-shirt featuring these fateful numbers.) In 1969, long from retirement, he surveyed his cheat sheets. "So we looked down at the list at that alarm," he said. "And if it doesn't reoccur too often, we're fine." What was the problem? The computer went through a cyclic period of calculations, repeating every two seconds. If at the end of two seconds, it hadn't completed its laundry list, it first complained with an error code. Next, it shook itself off and restarted itself—not a restart in modern standards, with a spinning color wheel or hourglass making a user nervous, but rather a carefully designed, instant reboot. Here was an enormous benefit of a simple computer. "Flush everything, clean it out, look at those restart tables," Garman explained. "And go back to the last known position and proceed forward." In each cycle, engineers had the computer running through the most important tasks first, so the tasks Apollo neglected on that July day were not critical to landing. The actual cause of the overload wouldn't be known for another day of round-the-clock, back-room work in Houston and at MIT's Instrumentation Lab.

Recordings of these tense seconds feature a reedy voice telling mission controllers "Go flight"—no need to abort the mission, keep going for the landing. Engineers mercilessly teased this youngster with falsetto renditions of his "Go flight" for months to come. But the chain-smoking engineers remained every bit as calm as, or even calmer than, the astronauts. The high-voiced engineer, Steve Bales, chalked it up to practice. "Looking back, the sims were almost as pressure-packed as the flights," he said. "Sometimes [they] would throw problems at us that were so hard. . . . We'd crash or abort the landing. . . . And we practiced right up until two days before the launch."[21]

The alarms kept popping up, since the computer was continually overloaded with information, and Houston watched pilot Neil Armstrong's heart rate climb to 150 beats per minute. As the lander approached the surface, the cycle time decreased as planned, because the computer sought to update itself more frequently. Now it finished even fewer of its tasks in one second, compared to two seconds, and the error code switched to "1201" versus "1202." Again, engineers advised Mission Control to ignore it.

In the end, the press portrayed this as a "computer error" overcome by the cunning of valiant humans. But the computer was actually just following orders. The human mistake, discovered after the landing, involved leaving *both* radar systems on. Even as the poor lander's brain tried to track the lunar surface below, it was also trying to find the rest of Apollo in orbit. Wouldn't the diligent engineers have tested for this? They did. But two things differed between simulations and the first real landing. For one, the docking radar, pointing upward, was now just feeding pure noise to the computer, since the rest of Apollo was nowhere to be found. Noisy input was much more challenging for the computer than sane, simulated input; noise chewed up more processing time. Next, engineers had made one tiny, sloppy error. In the Earth-bound labs, they hooked both radar systems to the same battery supply, unwittingly creating a false digital harmony. On the actual lander, each radar had a separate source of power. Since they were *not* happily synchronized in space, the staccato nature of their blather made the computer work twice as hard to listen. (Imagine two friends talking to you at the same time instead of taking turns.) If anything, the great design of Apollo's computer program—its jukebox-style "software"—saved the day. If it hadn't been so robustly designed to restart itself and brush aside such errors, the astronauts would have been forced to abort the mission.[22]

The landing was tense enough, even ignoring alarms. After the noisy confusion, astronaut Armstrong spent extra minutes looking for a good spot (i.e., no boulders, no slope). Given the best photos available, NASA had chosen what *looked* to be as smooth as a parking lot, but the intended site, once Armstrong took in a good eyeful, was strewn with car-sized boulders. With the lander still gliding too quickly to land, about thirty miles per hour and five hundred feet high, a new problem emerged: an empty fuel tank warning.

Marlowe Cassetti provides another window into the engineers' obsessive dedication. Cassetti, after all, had lugged a teletype home to his dining room table to make sure this day would work. "When they called out 'low level light,' I thought, 'Oh, my God. I did something wrong,'" he said. "'I didn't take everything into consideration.'" It turned out this low fuel warning was largely mistaken. Cassetti now labels it a "very interesting phenomenon." As the lander approached the Moon, it stopped braking, and physics no longer pushed the fuel against the bottom of its tank. The Moon's relatively weak gravity allowed the fuel to slosh around much more than it would on Earth. Hence the fuel gauge was increasingly inaccurate as the craft approached its landing. "It was a cliffhanger for me," Cassetti said. "Of course, everybody gave me hell for it. 'Marlowe, what did you do?'"[23]

But Armstrong did find a spot, avoiding boulders and craters. As the lander set gently down, Garman snapped out of a trance. Everything had felt like just another simulation until he heard an astronaut mention that their engine was kicking up lunar soil. He later said this was the moment when the magnitude of the milestone hit him.

Frank Hughes, the simulation master, stood with a colleague in Mission Control and they found themselves surprised. "Son of a bitch," said the coworker.

"Goddamn," said Hughes.

He later tried to describe their mindset. "None of us thought 11 would make it, I mean, just statistically, something had to go wrong." They didn't necessarily mean disaster, just an aborted landing and a return to the command module. After a handshake, Hughes and his colleague got back to work. A major checkpoint awaited, just one minute after landing. If the engineers found certain problems, they would have the astronauts immediately launch back to space. Another major checkpoint would follow, ten minutes after landing.[24]

Simultaneously elated and worried, NASA realized it wasn't sure exactly where the lander was. As astronauts described the local terrain viewed through their constricted windows, Mission Control tried to match those descriptions to existing lunar maps. One engineer recalled the scene as nearly comical. When they finally triangulated Apollo 11's historic spot, it was four miles from its intended target. This rare imprecision rankled the troops.[25]

In the constant earthly chatter, engineers found more reasons to avoid celebrating the milestone. To some, the lander's angle of repose was too great. Instead of being on absolutely flat ground, it listed slightly. Engineers wondered if it could take off safely from such an inclined position.

Another group, monitoring data streaming back from the Moon, drew sharp inhales. Something was wrong with one of the fuel lines. Temperature and pressure readings quickly spiked, and nobody knew why. There wasn't a lot of fuel left in that thin pipe, but when hot enough, it became unstable—they risked a deadly explosion shortly after this remarkable landing. One of the lander's designers recalled being suddenly drenched in sweat. "Nature took over and solved the problem for us," Tom Kelly wrote later. The line had become clogged by a chunk of frozen fuel. "Heat soaking back from the engine melted the fuel ice plug . . . and the pressure abruptly dropped to a low value. We looked at the screen in amazement for a few seconds, then broke into smiles and cheers of relief."[26]

The famous moments of astronaut Neil Armstrong descending a ladder to the lunar surface provided more breathless moments for engineers. One of von Braun's long-time German colleagues, Ernst Stuhlinger, recalled his own sense of responsibility. Early on, von Braun had charged him with divining the scientific consensus about whether or not the Moon's surface would support a lander and human footsteps. Stuhlinger had found that, overall, most thought it would work, but some still-dissenting scientists had said astronauts might encounter deep drifts of dust. Now an astronaut slowly descended the rickety, low-weight ladder. "We held our breath," Stuhlinger recalled. "Does he sink in or does he find solid ground?"[27]

Armstrong's last step was a significant drop to the surface, looking to many like a sudden lurch. The engineer who'd developed the last-minute stowaway flag gulped. "What went through my mind was the ladder broke," Moser said. "The sharp edge got his space suit, put a hole in the space suit, and the whole lunar program was over." Every engineer had a horrible projection of how his or her unique mistake could ruin everything.

Astronaut and pioneer Armstrong, the first human to set boot on another terrestrial body, then made his famous statement (either misspoken or, as he later maintained, misheard), intended to be "That's one small step for a man, and one giant leap for mankind." But the transmitted, transcribed, and forever-quoted version, making less sense, omitted the "a" before "man."

An enormous television audience on Earth took in these moments. Some, like Henry Pohl's wife watching at home with their children, worried that Armstrong could fall off the Moon, or just accidentally leap off it, given the strange gravity everyone was talking about. In League City, Texas, a friend of my parents snapped our first family portrait: a beaming engineer father sitting on the floor, and his wife holding an oblivious baby, with the blurry televised image behind us.

Audiences took for granted that the televised pictures were so poor, so grainy. After all, the signal traveled *all the way from the Moon*. But the televised images suffered a single bad decision in the early 1960s. Originally, engineers had just planned for radio signals. In designing Apollo's communication systems, they didn't leave a lot of extra room (in terms of signal bandwidth) for something as complicated as a TV transmission. To their credit, it may not have been clear to your average nose-to-grindstone engineer in 1961 what television would become by 1969. Even late in the game, some were against having a TV camera aboard Apollo. It was just something else to design, another heavy object, more astronaut training, and more equipment that could fail. The compromise was a specially crafted black-and-white signal carrying just a fourth of a standard TV signal's resolution. It also looked choppy because it updated the screen at a slower rate than normal TV. (Regular broadcasts updated screens sixty times per second, whereas the Apollo TV signal updated just twenty times per second.) In a final insult, the Moon signal wasn't compatible with standard earthly infrastructure. Engineers had to display it on a special system built for the Moon signal; then they pointed a normal television camera at *that* screen. Faget, who'd been a leading voice for televising the first Moon walk—how could we not, he argued, after spending $20 billion to get there—shook his head at the "stingy"-looking result.[28]

Still, audiences on Earth were spellbound. An estimated 95 percent of American televisions, and a total of one in six of the world's population, watched. Walter Cronkite had a rare moment as his eyes filled with tears. "I'm speechless," he said.

In fact, humans never have known how to digest something so completely alien, this miraculous feat of engineering and organization. What did it mean? What could anyone say about it? Many relied on a hackneyed analogy to Columbus sailing across the Atlantic and "discovering" the Americas. But those continents were already brimming with people. These new space vessels were nothing like creaking sailing ships, and colonies would not be sprouting upon this desolate orb.

Cronkite, a long-time second place finisher in nightly ratings, would soon leapfrog his NBC competition. He rallied from his tongue-tied moment to bring the discussion back to Earth, and eventually, in filling the long airtime between updates, he poked critics of NASA. "I'd like to know what those kids who were kinda pooh-poohing this thing are saying right at this moment." Despite this squares-versus-hippies narrative, most criticism for the Moon shot had risen from either research scientists (e.g., Apollo wasn't designed to actually discover much) or voices, like Reverend Abernathy's, from American social movements (e.g., space exploration seems like a foolish or even cruel way to spend money, given poverty at home).[29]

Even if they watched without milk and cookies, actual hippies did tune in and marvel. In one Houston backyard, a young group of radical architects and artists, including some hailing from California's "Ant Farm" collective, built a special viewing chamber for the event. They assembled an enclosed "space egg" of black plastic persistently inflated by an electric fan. The group gathered in their tent-like egg after dropping acid, and tripped with the television flicker providing the only light within. The squeaky, scratchy radio signals of astronauts mixed with the fan's hum. At the conclusion of the lunar transmission, the group had the brilliant-sounding idea to take the whole apparatus to the space center itself and assemble it there for all to see by morning's light. Re-inflating the egg in a gas station parking lot, they were surprised that nobody else was there to behold it. But as NASA and most of their audience well knew, the mission was only half complete.[30]

The two astronauts took a busy two-hour excursion outside the lander. They reported on the mobility and temperature of their suits. Armstrong took a call from President Nixon. Fellow astronaut Buzz Aldrin literally kicked lunar dirt as part of a planned "soil mechanics" experiment. They planted the American flag, collected about fifty modest Moon rocks, scooped a few pounds' worth of lunar soil, and—using their special Hasselblad camera—took about 150 photographs. Then they carefully climbed the ladder. Before repressurizing the thin balloon of the crew cabin, they unhooked their sustaining backpacks and, in mid-century fashion, tossed these and other depleted articles overboard as Moon litter. After closing the hatch, they settled down for a few hours of sleep before a planned launch from the surface. The lander's main designer left Mission Control and tried to get his own sleep in a nearby hotel. "But I kept waking up and worrying about a micrometeorite strike or window failure," Tom Kelly said, "leaving the crew gasping for breath in an airless cabin. I knew the odds were overwhelmingly against this, but that did not stop my worrying." The astronauts probably slept better than most of the engineers in those hours.[31]

When it came time for liftoff, the lander left behind its lower segment, legs and all. The crucial ascent engine worked like a champ, and the slight tilt of the lander created no trouble. The only casualty of the lunar liftoff was Tom Moser's flag. Planted too close to the lander, it fell over in the engine's exhaust.

The legless version of the lander rejoined the rest of the Apollo spacecraft, and all three astronauts gathered together in the command module, preparing for a two-day trip back to Earth. One of the only remarkable moments of their return came within the closed eyes of the astronauts. Space was truly not black. Stars of the Milky Way spilled across the heavens, and shafts of sunlight made hour-long orbits within the capsule as it spun in barbecue mode. Even then, instrument lights and indicators had their own patterns and rhythms around the astronauts. Shut-eye did not come easily, but orders were orders, and the crew attempted to sleep. Even with their eyes closed, space toyed with them. Astronauts later reported seeing bright flashes, at random intervals. Were they hallucinating or maybe just exhausted, with fried nerves?

Weeks later, when a particle physicist heard of this odd event, he ran to his local particle accelerator at Lawrence Berkeley Laboratories. He had predicted these flashes might occur but had never tested the idea. Worried only with discovery, he donned an opaque hood, closed his eyes, and lowered his usually brilliant head into the accelerator's beam line and the rush of heavy ions. Sure enough: flashes! Super-energetic (and most probably harmful) particles interacted with the eye's inner liquid, creating a cascade of light. The astronauts, even absent a solar flare, were unwitting targets of heavy cosmic rays. These relatively hefty, invisible bullets stream through space in all directions.

We still aren't *completely* sure where these heavy cosmic rays started. We believe they likely erupted from exploding stars and have been traveling about our galaxy for millions of years. The full health impact on our bodies is also murky; we know it can't be good, but the question is simply "How bad?" How much of a radiation dosage did the Apollo astronauts receive? According to Dr. Eugene Benton, a physicist who helped develop Apollo's radiation detectors, each astronaut wore three—chest, thigh, and ankle—and for a full Moon mission, these detectors each recorded a handful of heavy cosmic ray impacts. But since these rays can include, for instance, the guts of an iron atom moving at a good fraction the speed of light, they can do some damage. And if the little detectors each registered a few impacts, an astronaut's entire body endured hundreds or even thousands. One recent study shows that the astronauts who've left Earth's influence for several days have had surprisingly high rates of cardiovascular damage compared to earthbound humans.[iv] While we can debate the effect of a week-long trip to the Moon, the issue of heavy cosmic rays looms large for missions to Mars or beyond.[32]

Once the astronauts splashed down, celebrations erupted like never before on NASA Road 1. Henry Pohl just wanted to get home to his family but found himself bumper to bumper with happily honking cars and roving bands of revelers. Once he'd navigated home, Pohl loaded up his family and took off for his parents' south Texas ranch; it was time for fresher air. Even the cattle would give his nose a welcome and homespun break from days of cigarettes and stale coffee. He wanted to lie back and regard the night skies the old-fashioned way.

Huntsville, too, rejoiced. Von Braun rode on the shoulders of city leaders, up the courthouse steps for a not-quite-impromptu speech. Once the church bells and air-raid sirens quieted, he didn't make any promises but suggested, "Maybe one of these days we'll even have a man on Mars."

The astronauts had to delay any celebration. NASA quarantined them, just in case the mission had picked up a deadly microscopic stowaway. Faget's "receiving laboratory," meant to properly store and protect lunar material, now became a facility that also aimed to protect Earth from possibly harmful alien organisms. The extreme caution turned out to be overkill—the Moon, as expected, is completely lifeless. Some engineers found the caution ridiculous, but the price of being wrong was too great.

[iv] The paper did not account for an Apollo astronaut's hundreds of post-mission banquets with mandatory steak and dessert.

Once on the deck of an aircraft carrier, the astronauts boarded an Airstream, and the sleek trailer became their quarantined home for the next three weeks. They read books, scribbled their thoughts, and played board games with one another. As awkward as that might sound, especially for three humans who'd already spent a nonstop week together, these lodgings greatly expanded their space and comfort from the Apollo command module. Engineers and medical officials held various debriefing meetings with the astronauts through a glass wall. Doctors monitored the astronauts' vital signs every day until they were deemed pathogen-free. For additional weeks, many packets of film, Moon rocks, and other goodies sat like the world's most exciting Christmas presents until everyone from the medical community felt sure they were uninfected. Various animals (including oysters, houseflies, quail, and mice) were injected with lunar material and monitored for any signs of illness. (Agitated swimming of the planarians provided the only reaction to lunar dust, but who could really blame them?)[33]

With the primary mission accomplished, the engineers ramped up thoughts of "What's next?" But NASA had no collective, compelling answer. It seemed that the nation's new president, Richard Nixon, would curtail NASA's budget even more rapidly than Johnson had. Like so many fireworks lined up on the 3rd of July, another nine Apollo missions sat ready, pointing at the Moon, with increasingly more scientific work in each one. But what stretched beyond that? Some top administrators wanted to push for an actual Moon colony and a trip to Mars. Others, like von Braun, read the tea leaves and recommended a more modest and practical path, something that could at least sustain decent funding: an affordable Earth-orbiting laboratory, for one, and Faget's winged, reusable shuttle craft for another.

My father had transitioned into an advisory role by this time, focused on the future of mission planning. Did that mean the later Apollo missions, or immediately after Apollo, or well beyond? "All of the above," he says. He recalls a trip to Grumman's headquarters in New York, where they manufactured the lander. NASA asked Grumman to modify the lander's design to allow for stowing more Moon samples. Scientists were already making great demands of Moon rocks, so bringing back more made good sense. The trip went well, and Grumman agreed to lose one of the batteries, making room for more rocks.

Despite the triumph of the lander, Grumman, like other space-reliant businesses, housed some nervous employees after the Apollo 11. The majority of the space program's four hundred thousand people worked within one of fifty thousand different contracting companies. The five largest space contactors were all ranked within the top forty of the U.S. economy, each doing between two billion and three billion dollars of annual business. More than one hundred of these companies had prepared special press kits to promote their roles in the Moon shot, but with NASA budgets falling and the end of Apollo already in sight, many engineers approached abrupt changes.[34]

Most of NASA busily prepared the next Apollo launch, and by mid-November, another Saturn V rocket, with the next Moon-bound ship, sat ready to go. NASA sometimes rewarded engineers by flying them to Florida to witness a launch in person. Mac Henderson had spent his first years at the agency studying the first eight minutes of an Apollo launch, planning for every possible problem and contingency during a Saturn V's first two stages. What calamities would cause it to abort, pulling the crew capsule away from the massive tube of explosive fuel, and what calamities could be allowed to ride to orbit? But after several successful launches and with robust staffing in Mission Control, they rewarded this behind-the-lights engineer with a spot in the spectator stands that morning.

"I was a young fellow," Henderson says, and the idea of a launch was especially exciting. On launch day, clouds rolled into the Cape and thickened for a long stay, but NASA decided to proceed. The Saturn V roared aloft, as sound waves buffeted the crowd, and then disappeared into the low cloud cover. About a half-minute later, bright flashes lit the sky. "You couldn't see the vehicle," Henderson recalls. "But you could see the lightning." Uh oh. "Then the loudspeakers come on and they're saying, 'Hey, there's been a problem, we need y'all to leave.'" Despite no sign of lightning earlier that morning, the ionized exhaust from the Saturn rocket acted as a perfect lightning rod, inviting two strikes within a minute.

Staff quickly herded the assembled guests onto waiting buses. If the Apollo rocket was going to have a major problem, they wanted these people to get away from the launch complex. Henderson, the expert on early-launch troubles, felt helpless boarding a tourist bus. He was missing the one minute of the Apollo era that needed him most.[35]

In the command module, the astronauts rocketed onward with no electrical power. They deduced they'd been hit by lightning, but they were injury free. The engineers in Mission Control weren't sure what had happened and weren't sure what to do. They'd lost contact with Apollo in a burst of static. "The whole place just lit up," a young engineer named John Aaron reported. "I mean, *all* the [warning] lights came on. So instead of being aids to tell you what went wrong, the lights were absolutely no help at all." And then there was data like nothing the engineers had ever seen—Apollo's instruments sent a stream of gibberish to Mission Control.

Should they abort? The Apollo craft's guidance system had lost its bearings completely, but the Saturn rocket soldiered on, with its brain still awake and keeping the mission on track. As the engineers watched the mission gaining altitude on a textbook schedule, they'd trust the Saturn V and for now try to somehow bring sanity to Apollo's instruments.

Aaron quickly saw a pattern in the gibberish. Incredibly, he recalled this odd fingerprint from a trial run. He'd had no active role in that earlier day, in what was called a "pad test," but he obsessively observed. When the gibberish emerged, the only thing that had fixed it, in that test, was workers at the Cape resetting one modest switch. He sent word up the chain of command, but nobody recognized the "SCE" switch—not his Mission Control bosses, and once they regained contact with the crew, not the astronauts.[v] Aaron, all of twenty-three years old, carefully told them where to find it (there were hundreds of switches lining the Apollo command module); soon the Apollo crew compartment powered up and regained its senses.

Engineers had actually worried a great deal about lightning and had taken significant steps to protect their tall stacks of Saturn stages and spacecraft modules. But instead of focusing on the eight minutes of ascent, where there was little to be done, they had reasonably worried more about the *weeks* that an Apollo mission spent crawling to the launch pad and waiting through myriad tests and preparations. Engineers who focused on electrical systems look back on lightning as one of their toughest challenges. They installed a specially designed "amp trap" on the launch towers to help absorb direct hits, and though it's difficult to find in photos, thin wires hung above the waiting tower and rocket to help channel any lightning strike to the ground. Nonetheless, direct and damaging strikes hit more than one waiting mission; Florida's frequent thunderstorms could not restrain their electric curiosity for these strange and temporary towers rising from the flat landscape.[36]

Once Apollo 12 had safely orbited the Earth, NASA took stock. They'd lost a couple of temperature sensors, but everything else seemed to be working. Lightning had sent enough current to power a small city through the rocket and parts of the vehicle, but Saturn and the tough little spaceship had shaken it off. Tape-based computer memory would have been fried, but with its hard-wired version, Apollo's computer program was unruffled. When looking back from our modern bureaucracies, the next step feels inconceivable. Their ship had lost power, and they had nearly aborted the liftoff after a lightning strike. Yet, NASA considered completing a trip to the Moon and back. In its typical fashion, the agency took individual delegations of duty as gospel, and the flight director of this mission, Gerry Griffin, would make the decision.

"We were allowed to do what we were good at," he said. But there were massive PR considerations, with congressional funding at risk if the mission failed, right? "We had checks, sure. But nobody from headquarters was trying to second-guess." The engineers would look over the data and make the most reasonable decision,

---

[v] SCE stood for "signal conditioning equipment."

measuring the mission goals versus the instrument readings and possible risks. After about an hour of collecting information from each system and subsystem, the decision seemed obvious. "We finally figured out we had to do a re-entry anyway," he says, noting the most dangerous part of any mission after the launch. "So, let's go to the Moon."[37]

Compared to the first landing, the second one was smooth and simple. Initially, when the lander departed for descent, the astronauts reported it was bucking like a wild horse—the thing just seemed incredibly sensitive. This turned out to be a slight simulation problem. The lander's simulation program had trained the astronauts for a much more sluggish craft. But soon, they had the hang of the real thing, and the guidance computer used a new and improved descent technique. Not only did it turn off the docking radar this time (focusing purely on the landing) but the computer also now compared the speed at each time step to what calculations had predicted. It then made subtle corrections along the way, and Apollo 12 stuck its landing on target.[38]

My father has always kept a framed black-and-white photograph featured prominently in my parents' home. "This is my favorite," he says. "It tells my whole NASA story." With the lander in the background, the photograph shows astronaut Charles Conrad Jr. from Apollo 12 standing near an unmanned Surveyor lander built by Hughes Aircraft, where my father had worked to put the first American hardware on the Moon. The astronauts took careful readings from the Surveyor's skin. Its subtle pock-marks gave the best measure yet for the frequency and size of micro-meteoroids striking the Moon. The "old" probe (it had landed two years prior) showed some minor space weathering, but overall, its lack of major blemishes suggested that the Moon's surface was a relatively safe place (see Figure 12.3).

With the third and fourth set of human boots on the Moon, the astronauts prepared to deploy a newly designed television camera. Now the audiences at home would get a properly crisp picture of Moon adventures, with the red, white, and blue of the American flag, and gold-colored foil crinkled around the lander. However, when struggling to set it up, an astronaut accidentally pointed its sensitive electric eye at the unfiltered sun, rendering it blind in an instant. Networks scrambled with back-up plans. NBC had studio actors in astronaut garb pace out what should be happening on this second Moon walk. CBS, following on their leading coverage of the first Moon shot, had a more elaborate back-up. Using a grainy camera, fake moonscape, and marionette astronauts, they simulated "live" footage of the excursion. The footage was fairly convincing, and some viewers assumed they were watching the real thing.[39]

This brings up an uncomfortable point. At a time when special effects were taking a great leap forward and when many Americans had started to question technocracy,

FIGURE 12.3 Fancy meeting you here. The unmanned Surveyor 3 probe receives a visitor, astronaut Charles Conrad Jr., with Apollo 12's lunar module in the background. (NASA photograph.)

charges of the landings being faked were somewhat predictable. Many people today might assume such conspiracy theories to be a contemporary phenomenon. Perhaps they feed on our era's unique psychology, buoyed by the internet, growing in number once an event recedes into the past. But in fact, doubts were widespread in 1969. Engineer Wesley Ratcliff recalls gathering family and friends at his home for the landings. "We had so many people tell me we were wasting our time," he says now. "And that it was just a big production." Likewise, Huntsville scientist Tom Parnell tells a story of traveling late one night by car, not far from Huntsville. He pulled into a rural filling station and saw television coverage from one of the Moon landings. He watched some of it there in the station office, waiting to pay for gas. The attendant walked in and looked at the screen. "You don't think they're really up there," the attendant said, "do you?"[40]

I've always known that my father's own mother never believed in the missions. No matter that her son worked long hours at NASA; this woman of the swamplands knew that the Moon was part of heaven and mortal men just could not go there. She passed away a few years before the landing. I long assumed this poignant story was somewhat unique to our family, but it wasn't uncommon at all. Many Apollo engineers had extended family members, especially from older generations, who just refused to believe we could step on the Moon. The big screen scenes of *2001* looked better than NASA's grainy, flickering footage anyway, so why couldn't it all be faked? It's easy to imagine a doubter from late 1969, rolling their eyes. *Oh, you pointed the new, better camera at the sun, did you? Mm-hmm. What a convenient story.*

The Apollo 12 mission returned safely on schedule, but NASA had watched American interest plummet. Camera snafus aside, the launch coincided with more bad news: the breaking story of the My Lai Massacre in Vietnam, with its shocking photographs taken much closer to home. As if American soldiers killing unarmed civilians wasn't shocking enough, audiences watched another emerging hostility, as a new White House administration openly pointed fingers at specific segments of society. Earlier in November, Vice President Spiro Agnew had attacked "an effete corps of intellectual snobs," including professors, hippies, and war protestors. The media, too, were just a "small and unelected elite," he said, at odds with real America.[41]

After a fleeting moment of unity, humans seemed to return to a natural, tribal outlook as they digested NASA's achievement. Counterculture writer Norman Mailer saw 1969 as the year his team definitively lost their war with the squares. He wrote that his kind were an "abominable army." He wrote that he wanted to scream at his bar mate, "You've been drunk all summer, and *they* have taken the moon!" On the other end of the spectrum, archconservative Ayn Rand simultaneously wrote that "Apollo and Dionysus represent the fundamental conflict of our age." She saw these not as "floating abstractions" at war, but as concrete and contrasting ambitions: one a technological temple at the Cape, and the other writhing in drug-fueled lust at Woodstock, New York.

The counterculture itself was fragmenting by December of 1969. After sex, drugs, and rock 'n' roll had offered a collaborative vision in Woodstock, version 2.0 fell to violent mayhem at Altamont Pass near San Francisco. With bad trips, dozens of beatings, a drowning, and a knifing victim dragged before the horrified Rolling Stones, many of the three hundred thousand in attendance left wondering if the years ahead would have room for peace, love, or moonbeams.[42]

Republican minority leader Everett Dirksen, a veteran of World War I, was called the "Wizard of Ooze" for his memorable oratory. Late in 1969, he proposed to his congressional colleagues a wholly positive and uniting gesture: The nation, he said, should henceforth recognize each July 20th as a national holiday. America should celebrate what was arguably the century's signature achievement. But his "Moon Day" idea failed to excite other legislators, and it disappeared without a vote.[43]

13

1972—FROM ROVERS TO REGRETS

Marlowe Cassetti recalls standing in his yard during an especially stressful Apollo mission. A neighbor had been watching the tense updates on television, and he asked Marlowe, "What are you going to do if this doesn't work?"

"I tried to keep a positive outlook," Cassetti says, but he didn't lie, and NASA had been open with the public about the mission's peril. He stepped his neighbor through the mission's status and its narrow path to survival. The neighbor shook his head. "Marlowe, you don't make it sound too good."

Apollo 13 was a near tragedy and arguably the most famous of all NASA's adventures. To a greater extent than most other engineers, Cassetti understood *exactly* how much oxygen and how much electrical power a wounded spacecraft would need to survive. "When I was looking over the numbers and talked to the flight controllers," he says now, "it was bewildering. I thought there's a good chance we're not going to get these guys home alive. Of course, I never said that out loud."

For a brief time, the imperiled mission brought American viewers back to Apollo, if only for the drama of three men in space, fighting long odds and relying on round-the-clock problem-solving back home.

In fact, the entire world followed Apollo 13. If America's enthusiasm had waned after the first landing, interest remained fervent abroad. NASA took Moon rocks, some spacecraft, and astronauts on a world tour, drawing incredible crowds at every stop. At the 1970 World Expo in Japan, people waited up to seven hours, forming a line a half-mile long, just to see a Moon rock behind Plexiglas, along with the Apollo 8 capsule that had trekked half a million miles in space.

At home, NASA set up an Apollo display on wheels, and it embarked on a fifty-state, one-year tour. This extended van raised side wings to display exhibits; by the end of its journey, more than three million Americans had strolled past to view various program artifacts. Yet, even a year after the first landing, Apollo was rocketing to its end. Due to funding cuts, NASA elected to eliminate the last three planned missions, leaving five flights to span 1970–1972. Many engineers were shocked when they heard that mission 17 would be the last.

In preparing for Apollo 13 in early 1970, technicians had assembled the mule-like service module. Among other supplies, it carried two tanks of liquid oxygen. These provided astronauts their breathing sustenance and fed the fuel cells to generate electricity for the Moon missions. After running various tests to make sure the tanks were leakproof, a technician's grip slipped and one medicine-ball-sized tank dropped to the floor with a loud plunk. It fell just two inches and looked fine, so they continued their work.

Whatever injuries the tank may or may not have suffered, it was definitely cursed. And stepping through its story provides a glimpse into the incredible layers of Apollo's technical detail, each fraught with potential disasters.

Once snug in the service module at Cape Canaveral, the tank showed itself to be stubborn. Technicians filled it and then emptied it as part of routine preparations. But the tank refused to lose its last few gallons of liquid oxygen. Wasn't this a bad sign? NASA officials had to make a devil's choice—one of many. If they started tearing apart the oxygen system, that could delay their launch. But more importantly, extra tinkering with the service module could damage *other* equipment, including the second oxygen tank. Engineers thought the tank's problem was probably a slightly loose interior tube, and on its own, that shouldn't jeopardize a mission. NASA decided to keep this oxygen tank and march onward. To fully empty it, the engineers used an interior heater to simply boil any residual liquid oxygen away.

But now an insidious detail, lurking in deep technical weeds, bit the agency. The Cape had upgraded launch pad electrical lines to triple their original voltage. And while NASA had sent this detail along to the tank manufacturer, the numbers were buried with countless other myopic specs. In short (electrical pun regretted) engineers force-fed the tank's heater three times the voltage it expected.[i] As technicians warmed the tank, they let the heater run about eight hours. The tank's little thermostat couldn't handle the new excess voltage, and it fused into a permanently "on" position. Without alerting anyone, the tank rose to a temperature of about 800° Fahrenheit, melting some of the wires' insulation. None the wiser, engineers refilled the empty tank before launch, and Apollo 13 carried a broken tank, with newly bared wires immersed in liquid oxygen, into space.[1]

The astronauts completed their first maneuvers without incident. They pointed themselves on a path to the Moon and twirled barbecue style for the next two days. Eventually, with the Moon growing larger in the capsule windows, an astronaut marched through his checklist and hit a switch to stir the oxygen tank: routine business in the realm of zero gravity. But this jolt of current through the tank's damaged

---

[i] The electrical *mode* had apparently changed as well—alternating current versus direct current.

wires cooked up a small fire. Nearly two hundred thousand miles away from Earth, the astronauts heard a loud bang, like nothing they'd heard before from control thrusters or simulations or anything else. They'd actually heard the oxygen tank explode.

The night shift had started in Houston. Engineers tried to sift through very odd data coming from the Apollo instruments and the reports from confused astronauts. It looked like they were rapidly losing their oxygen supply. "First of all, we thought we'd boil it down to something simple and obvious," engineer Arnold Aldridge recalled later. "As it went along, it became clear it was more extensive." He made a phone call to John Aaron, the youngster who'd seen a pattern in the data noise after Apollo 12's lightning strike.[2]

Aaron recalls being "at home, shaving . . . after spending a long shift at the Control Center." The new shift updated him rapid-fire. They thought they were facing an instrument problem or "flaky readouts"—there was no way the disturbing data could be real. Aaron asked to hear the numbers from various systems, one at a time, over the phone. "That's not an instrumentation problem," he told them. "It's a real problem." Thinking back on it, he saw his distance from the Control Center as good fortune. He could stand there calmly in his bedroom and try to see the entire forest. "Everyone was glued to his own tube, and he was digging into his own area deeper and deeper and deeper." But Aaron thought Occam's razor was showing them something sinister in the simplest explanation. "You guys are wasting your time," he told them. "You really need to understand that the [spacecraft] is dying."[3]

Once they believed the readouts, the engineers immediately started thinking through the problems. They'd already lost most of the stored oxygen in the service module. Chris Kraft had moved into more of a managerial role in Apollo, but he returned to the trenches now. Kraft realized that those tanks were connected to the three smaller oxygen tanks in the astronauts' command module. In happier times, those tanks kept a nice supply of oxygen near the crew, but now it meant the awful opposite: they were surely draining backward through their ruptured sibling. He called out, "Seal 'em off," and saved precious gallons of oxygen. But the prospects remained grim. In the words of one engineer, if they had seen such a scenario during a practice mission, "we'd have said, 'Well, you can kiss those guys goodbye.'"

As word spread through the ranks by phone, everyone who could possibly help swarmed to the center. Aaron got dressed and drove in. Aldo Bordano, in his early twenties, remembered the phone call. "It was 11:30 or something like that," he said. He drove the ten or so miles back to the center. "All of a sudden, it was about midnight, it was just a line of cars with their lights on. And that was us. We'd all gotten called in—all three shifts." He remembers the cars all filing efficiently into the parking lot outside Mission Control and Building 30. "It was a real eerie moment," he said. "There might have been two hundred of us turning our cars off and walking in at the same pace. Nobody said a word to the guy on the left or the guy on the right. We just went to our stations."[4]

The Apollo craft, kicked by the bursting tank, was off course as it approached the Moon. "We decided we had to make a burn real quick because we were going out into space," Henry Pohl recalled. "And we needed to . . . get on a free return back." The free return was the ideal figure-eight path. It would glide around the back of the Moon, quicken with the Moon's extra pull, and sling back toward Earth. "At that time, when we made that decision, [it] didn't look like it was a very good option to bring the crew back alive," Pohl said. "But it looked like it was a very good option to bring the crew back. You know, at least if they didn't make it alive, at least they wouldn't go off and be lost in space forever."

Scores of engineers started calculating backward. John Aaron became the czar of electricity. With the fuel cells mostly useless now, the engineers had to conserve every bit of electrical power. They knew how much they would need, at minimum, to get the crew capsule to re-enter Earth's atmosphere at the proper angle. Aaron meticulously worked from that goal, backward, to every minute of the return path, ruthlessly budgeting electricity, and sometimes saying a firm "no" to engineers more than twice his age who wanted more juice for their particular sub-systems. "So we had to cut our energy consumption in half in order to make it back home," Pohl recalled. "Well, to do that, you've got to turn every heater off that you don't absolutely have to have." As dramatized in reenactments, the spacecraft was going to get cold, very cold. But freezing the crew wasn't the only worry.

Like Aaron, Pohl also now had to work backward. The Apollo's little thruster rockets were one of his main responsibilities. "I calculated . . . how cold the [thrusters] were going to get, and I gave myself four degrees above freezing on it," he said, recalling his small margin for error. If the propellants froze, it wouldn't matter how much electrical power the astronauts had at their disposal, because they wouldn't be able to maneuver. They would be close enough to see the welcoming Pacific Ocean from space but unable to negotiate re-entry, either burning up or bouncing off the atmosphere and sailing helplessly away. In the end, Pohl says his propellant almost froze—it was only two degrees Fahrenheit away from disaster. "We cut those margins pretty dad-gum close."[5]

NASA had the astronauts move into the lander for most of the journey, since it had working batteries and its own oxygen supply. While it was designed for all things lunar, this time it drove the whole bus home. Its main engine (designed for landing) and its various thrusters (used for docking and fine adjustments) were fully functional. Even though it wasn't designed to move around with the rest of Apollo attached—like an acrobat balancing a stack of plates on his head—it worked well enough. And while this is sometimes portrayed as a brilliant last-minute idea, the engineers had game-planned it in advance. It was one of hundreds of practiced horror stories. The lander's designers had thought about using it as a "lifeboat" from their first musings in 1961. Allowing such a mode required more fuel and oxygen for the lander. But in turn, that provided a fringe benefit: A healthy mission could linger longer on the Moon. And in 1963, the company building the other modules asked if the lander's thrusters might be able to maneuver all the modules together, in a worst-case scenario where the command and service modules lost power.

Engineer Cynthia Wells (she who had persevered in those all-male math classes) recalled a 1967 NASA assignment: a lifeboat study. In an emergency, how could the lander basically run an entire mission? "Everybody laughed at it," she said. "Because everyone thought it was so stupid. . . . The flight controllers thought you'd never need to do that." But just in case, her group calculated, for instance, how long the lander could fully pressurize air in the command module.[6]

The unlikely had become the only option, and now little details—some helpful, others alarming—came home to roost like never before. With the lander functioning as a lifeboat, it was greatly increasing its planned occupancy time, and engineers wanted the astronauts to snag some extra carbon dioxide filters from the command module. This wasn't an optional activity to spruce up the air—too much carbon dioxide would be lethal. But with a slap of their foreheads, engineers realized a mismatch: the two modules featured carbon dioxide filters of incompatible shapes. Using only what was available to the astronauts, engineers in Houston figured out how to make an ugly sort of adapter from cardboard and tape, solving a literal square-peg versus round-hole problem.

One step omitted in most dramatizations, however, involved secretaries volunteering for "duties as assigned." Engineers were worried about describing improvised solutions to tired, shivering astronauts, with tasks well beyond their training. The astronauts weren't engineers, they'd have no diagrams to consult, and they might get too confused or frustrated to complete the tasks. Engineers placed a secretary who'd never worked on the modules in a room with a pile of Apollo-like supplies, including filters, cardboard, and tape. And then they relayed instructions, describing the fix via phone. After it worked smoothly in Houston, the engineers decided the directions were clear enough to try on the astronauts.[7]

With the clock ticking in the lander, mission planners looked for options to hasten the return journey. They decided to burn the lander's main engine, speeding the whole Apollo chain on the far side of the Moon. Authors Murray and Cox describe having the lander trying to push the other modules as "using a small car to push a limousine, but in three dimensions." (The weight of the combined command and service modules was about twice that of the lander.) On the back side of the Moon, out of communication with Earth, the astronauts started the requested extra burn of the engine, running it for about four long minutes. Once the mission emerged, engineers felt their hopes rising—the burn looked like it had worked perfectly and the astronauts just might make it home. They'd be miserable and cold, to be sure, but they could have just enough air to breathe.[8]

An animated Max Faget then made a rare entrance into the Mission Control scrum and sparked an argument: It boiled down to engineering versus human physiology. The astronauts were dog tired, having rested little since the start of the crisis.[ii] Many engineers wanted them to sleep for a few hours, but Faget was adamant. No, he said, Apollo had to enter barbecue mode as soon as possible, no matter the difficulty. (Normally, they used the command and service module thrusters to set the spin. Using the lander's thrusters posed a new challenge.) Every hour Apollo sailed home without a good spin meant an hour some component facing the sun could be melting and some fuel line facing the void could be freezing. His argument won the day, and the exhausted astronauts, on their second convoluted attempt, coaxed the lander into spinning the wounded Apollo chain into an acceptable, steady rotation.

With no power to spare for cabin comfort, astronauts, with no blankets, suffered plunging temperatures in the lander. Records show a low of 43° Fahrenheit, while the abandoned command module sunk below 40° Fahrenheit. In addition, to avoid any nudge off their homeward path, engineers asked the astronauts not to make any waste dumps. So, they struggled to find new ways to contain their urine, filling every spare plastic bag from either module. The discomfort of their cold confines, with floating urine bags, combined with prioritizing water for equipment, meant the astronauts slowly became dehydrated as well. One developed a kidney infection.

The entire planet followed the journey home. A dozen or more foreign governments offered their assistance with the coming ocean recovery. In Mission Control, and at Grumman headquarters, engineers took catnaps on top of desks and conference tables, ignoring their managers' appeals to please go home.

Approaching Earth, the astronauts crawled from the lander to the command module, where they had to wake every system from a chilly and unplanned sleep. Would Pohl's thrusters work correctly? Would the computer start up? Were the parachutes frozen stuck? Preparing for re-entry, they jettisoned their plucky lifeboat, the lunar module. The astronauts' last call to Houston before entering the atmosphere thanked the engineers for their incredible work in getting them this far. After tense minutes of radio silence and the incredible (probably welcome) heat of re-entry, Apollo 13 gently splashed down. When all three astronauts emerged, the delirious engineers shouted in celebration.[9]

[ii] Jim Lovell, Jack Swigert, and Fred Haise.

Overall, there was surprisingly little finger pointing in the wake of the mission. A stack of small errors had caused a critical tank to explode in flight. Mainly, everyone at NASA felt fortunate. The tank could have easily blown just two days later, with the lander and two astronauts on the Moon and a lone astronaut left in a quickly expiring, powerless ship in lunar orbit. The mission's lifeboat would have been too far away to help, and they probably would never have made it back alive. As it was, the engineers had just enough time to work the myriad, entangled problems and get the crew home.

Behind the scenes, one could hear a bit of the inter-center tensions emerge when discussing Apollo 13. Each center had its own culture and its own way of doing things, and sometimes they shook their heads at one another. Engineers in Houston and Huntsville had long arguments with personnel at the Cape about their obsessive pre-launch tests. Guy Thibodaux, a longtime friend of Faget's, recalled his Houston-based view of the Cape. They "always wanted to run tests on everything," he said, referring to things like repeatedly filling and emptying the oxygen tanks while the rocket waited on the launch pad. He blamed the tank explosion directly on what he considered a ridiculous testing regimen. While the tests gave engineers confidence in those systems on the ground, the tests simultaneously could be wearing or degrading crucial parts and wires.[10]

The major television networks had been ignoring the routine transmissions from this, ho-hum, third trip to land on the Moon. But the emergency became a multi-day, high-stakes space drama—something the world had never seen. The mission made for a gripping, life-or-death story, racing the twin clocks of oxygen supply and electrical power, with plummeting cabin temperatures aboard a disabled spaceship. The triumph of Apollo 13 highlighted the work of engineers for the general public like no other mission. With the astronauts mainly just shivering for a few days, broadcasts had to at least *attempt* covering technical challenges, clever emergency fixes, and the years of careful planning that had paid off.

Yet a few, like Marlowe Cassetti, have always wondered if it hurt the space program. "I think there was a whole mood that changed in Washington," Cassetti said. "Apollo 13 scared the management and they thought it was way too risky."[11] The public too was getting the idea that space was less romantic and less full of wonder than they'd assumed in 1958. Space seemed mostly colorless, incredibly dangerous, and either boiling hot or deathly cold by instants. Spaceships were not sleek; they were claustrophobic and uncomfortable. Life in space was razor stubble, messy hair, urine bags, and nausea. And even in a nail-biter like Apollo 13, journalists and audiences had to wade through drifts of technical terms and acronyms.

When Hollywood's *Apollo 13* movie came out in 1995 (the era of VCRs), my father received about five copies as gifts. He was always appreciative, but we couldn't convince him to watch it. "I don't understand why I would want to watch this," he said. "I lived it, and nothing could match that." But in retrospect, Apollo 13 gave children like me a handhold for appreciating and even bragging about our NASA engineer parents. "You know Apollo 13?" we could tell our friends. "My dad was one of those pocket-protector and horn-rimmed glasses-wearing dudes figuring things out."

The last four flights to the Moon, all with successful landings, spanned January 1971 through December 1972. Having summited the engineering mountain, NASA could focus on lunar science. This had long been the plan, from the earliest discussions, as engineering-dominated Apollo tried to appease the scientific community.

Nine months after Apollo 13's near disaster, the fourteenth mission flew with not only a redesigned oxygen tank but also an extra tank, extending the possibilities for the length and ambition of the mission. Apollo 14, with much less fanfare, came very close to aborting its landing.

By this time, the control center had enjoyed an upgrade. "Rather than trying to read all the actual digits," computer expert Jack Garman said, "you just had these banks of lights all over your console—I had hundreds of them—of every significant event or item or error." Green meant good, yellow meant caution, and red meant bad. The little lights had labels, but the engineers learned to read the patterns of twinkling lights as geometry, a sort of dynamic cuneiform. What they saw in Apollo 14 was an abort alert that would come on and then go off. They would ask the astronauts to reset a switch and breathe a sigh of relief when the abort light went off, only to see it come on again. Astronauts even resorted to an old-fashioned fix-it method: banging a fist against a console full of switches. This often worked, but only briefly. Engineers on Earth had to solve this problem quickly or cancel the landing. It was "the worst nightmare of all," said Garman, who'd notably already been through Apollo 13. "I looked around, and standing behind me were about ten people. Every icon of the space program . . . I mean all of them: Gilruth, Kraft, all of them," he said. "I woke up that we were in serious trouble at this point. . . . We only had like two hours [of fuel]. . . . And worst of all, we had too many solutions." Dozens of different failed components, in either the wiring or in the computer program itself, could be causing the problem.

The fact that fist pounding often erased the alarm actually helped everyone figure out the culprit: a bit of loose solder within a sealed box of electronics. Without gravity to corral it, the little metallic pebble floated from spot to spot, like a drunk moving between nearby bars, connecting critical circuit paths at random. But astronauts had little time and none of the equipment to open up a circuit near the Moon.

With just minutes to spare, the solution emerged from a computer expert in MIT's Instrumentation Lab. He called Houston and gave them the commands for

Apollo's unique user interface. "Verb 25 noun 7 enter; 105 enter; 400 enter; 0 enter." While it was impossible to *reprogram* the Apollo computers, given software forged of magnetic rings and thick wire braids, he had devised a way—an early "hack"—to convince the computer to ignore its compromised abort warning. The third lunar landing could move ahead.[12]

Apollo 14 did include a not-quite-welcome sign of the early-1970s times. An astronaut, freelancing from his appointed tasks, agreed to run a number of extra-sensory perception "experiments" with some questionable non-NASA "investigators" on Earth.[iii] At appointed times, the astronaut concentrated his thoughts and tried sending psychic messages to would-be "receivers" at home. After four such attempts, the results were sadly discouraging. Luckily, astronauts still had the radio antenna.[13]

The last three Apollo missions ferried science packages built to run experiments on the lunar surface for years to come. To an untrained eye, the science spreads resembled the worst stereotype of a rural, southern front yard. When fully unpacked and deployed, the experiments radiated out from a central unit that looked something like an abandoned dishwasher, with an array of separate experiments extending via electrical tentacles across a lunar patch as large as a football field. Special radiation and ion detectors put their thumbs out to test the "solar wind." Seismometers could measure tiny vibrations in the moon, discovering all sorts of "Moonquakes." Some packs included thumpers that would make their own vibrations and listen for the Moon's reaction, like a doctor tapping a patient's chest. Other sensors measured the way heat flowed through the Moon's surface or probed its magnetic properties (see Figure 13.1).

Each of the last missions also hauled extra scientific equipment in the service module so that it could collect measurements while the lander was away. These extra pieces included detailed light measuring equipment, cameras for surveying the Moon, and lasers trained on the surface, to better measure its contours, including ways the Moon might flex and shift over time.[14]

The Apollo missions marked an abrupt pivot for the field of geology. Tools long developed for studying Earth turned excitedly to another body. The later missions brought back more and more material, from a greater variety of formations and terrain. A receiving lab staffer recalled understanding the shift for geology and astronomy both, when a visiting scientist had his first chance to cut into a Moon rock. "I looked over at him, and he's shaking, literally shaking. His face is gray, and the sweat is just pouring off his forehead," he recalled. "Here's a man who had devoted his life to cosmic materials, meteorites and everything else. He's looking at this sample from the Moon . . . and he's almost beside himself."[15]

[iii] Astronaut Edgar Mitchell.

FIGURE 13.1 Astronauts in Florida practice deploying a lunar science package in 1970. (NASA photograph.)

In addition to bringing back samples from the lunar surface, the astronauts in the later missions also made forays into scientific observation. Starting in the earliest days of manned spaceflight, the astronauts had endured an awkward and even frustrated relationship with science. Trained test pilots felt bogged down with the tedium of running senseless-looking experiments. Early on, they endured scientists raising all sorts of medical worries for humans in space. Then, when astronauts returned triumphantly from the Moon, scientists had insisted on caging them in a trailer for a few weeks. An early NASA astronaut once interrupted a meeting about new experiments to declare, "To hell with the scientific community!" Another eventually called science "a parasite" on Apollo.[16]

Still, many of the later Apollo astronauts relished their role as humanity's first *real* field geologists. NASA even hired a couple of research scientists into the astronaut corps, starting in the late 1960s. Harrison Schmitt, a geologist, flew on Apollo 17, the final trip to the Moon. Joe Allen, a physicist, worked as the scientific liaison to the Apollo 15 crew,[iv] the first to range far from the landing site to make geological observations and collect samples. Allen later just shrugged at the tensions related to science, and saw two different agendas necessarily competing for the astronauts' Moon time. "If you're a flight controller," he said. "You just want [the astronaut] to

---

[iv] On Apollo 15, Allen technically served as the capsule communicator to the astronauts during the Moon walks.

step down . . . and then come back, because this is a very risky place, and the more time you spend . . . the more chance there is for a catastrophe." However, a scientist wants the opposite: More time would mean more data and more chances for fundamental discovery. In training, Allen and his fellow astronauts embarked on a number of geological field trips. They studied rare, lunar-relevant rocks in rural Minnesota, mountain ranges in arid Mexico, and cratering effects on a Nevada bombing range. Geologists wanted astronauts to recognize, for instance, the difference between impact craters and explosive craters.[17]

To aid the astronauts as field geologists, the last three missions employed a last-minute device to increase the range of exploration: the lunar rover, developed by the engineers in Huntsville. A four-hundred-pound electric cart meant to carry rocks and astronauts, the scrawny rover could not support a person in Earth's gravity, but it zipped nicely about the Moon. The rover contained some remarkable engineering. As rubber tires were heavy and ill-suited for a vacuum, engineers built tire treads from a mesh of zinc-plated piano wires. Even as the rovers provided much greater range to lunar exploration, engineers had to worry about the astronauts getting lost in such a colorless and unfamiliar terrain. And given the Moon's paltry magnetic field, a compass would be of no use. Each rover had its own guidance system that could track how far it had moved and which ways it had turned. As a remedial backup, they also featured little sundials to help astronauts orient themselves if all else failed.

The fifteenth mission was a sort of high point in terms of NASA's lunar television signal. The color picture now sparkled with striking quality. Networks aired special-interest segments explaining the rover and the geology focus of the mission. Even when some of these segments went wrong, they made for good viewing. Astronaut and physicist Allen recalled NBC wanting to show off NASA's new lunar drill, designed to extract samples from twenty feet under the surface. On live TV, in a little sandbox that NBC provided for their demonstration, an astronaut punched the go button and suddenly sank to the studio floor. "Everyone thought the drill stem had somehow broken," Allen recalled, laughing. "What had really happened, it had drilled down through everything . . . the steel . . . into the concrete . . . and had come out the ceiling below, all in a matter of about five seconds."[18]

Apollo 15 marked the last mission to have long, live broadcasts carried on the three major networks. This time, NASA cleverly left a camera behind to record the liftoff, as the lunar module rocketed away from its own leggy landing platform and up into the Moon's starry sky.[19]

Simulations expert Frank Hughes recalls something else memorable about the crisp movies of the rovers. "Anybody who didn't believe we went to the Moon should look at those," he said. "When the dust would fly [from the rover's wheels], it would fall flat. It didn't hang in the air. . . . There was no dust hanging in the air or anything like that." Dust and dirt kicked up by the rover fell in nice, perfect parabolas, sans floating or drifting. Such an effect would be impossible to create in an air-filled earthly studio. Hughes says he used to share this and countless other bits of logic with Moon landing skeptics but, as so many other Apollo engineers have, "I just gave up."[20]

Planners understood that the astronauts would need greater physical mobility in these last, more active missions. Seeking a suit for rock-hauling astronauts, NASA opened a competition between various suit makers. The Latex Corporation faced stiff (sorry) competition from a hard-shelled suit, but they submitted a crucial component in their pitch for their "omega" suit. In 1968, they had loaded a station wagon with recording equipment and transported a omega-suit-wearing coworker to a nearby football field. Over the next few hours, they had him do everything a football player does. He ran around with a football, threw a decent pass to another employee, caught passes in return. He even punted with his astronaut boot. The film of this odd scene, submitted to NASA, sealed the deal, and the Latex Corporation got the contract.[21]

In all, the Apollo missions brought back about 2,400 samples of different lunar material, weighing about 840 pounds on Earth. The lunar receiving laboratory, long fostered by Faget against various funding threats, moved quickly from a quarantine facility to one of preservation. Most lunar rocks, three billion years old on average, own a pristine nature due to the lack of atmospheric interference—no wind, rain, or even oxygen to blemish them. As of Apollo 15, the laboratory shifted from making sure nothing could get out to making sure nothing from Earth could sneak in to contaminate these invaluable samples. "I'm talking about a very, very small difference," the lab's director explained of the 1971 shift in pressure. Now air gently pushed out of the lab, discouraging contamination from the outside, where it had previously sucked in, discouraging escaped material.[22]

The last three missions, with their rovers, accounted for 75 percent of Apollo's total rock haul. But even the first rock from Apollo 11 had immediately shaken the field of planetary science.

The receiving lab included a special underground chamber for measuring the natural radioactive fingerprint of each lunar sample. Twenty-five feet underground and freed from the normal radiation noise at ground level, scientists could measure the samples in relative quiet. Many leading scientists expected radioisotope dating to find these Moon rocks, collected from the dark *maria* regions, to be 300 million to 500 million years old. The *maria* were relatively clean and smooth. Because they featured much less cratering than the lighter, highland segments of the Moon, conventional wisdom suggested these regions had formed relatively recently in astronomical terms. In 1969, once the first Moon rock cleared quarantine, scientists eagerly descended to their deep, monastic laboratory. "That first sample blew all these theories out of the water," says modern-day lunar scientist Caleb Fassett. At 3.7 *billion* years, if this sample wasn't a senior citizen in the solar system, it was headed for retirement. Suddenly, scientists had to accept what the Moon was telling them: The most violent times for the solar system, with huge impacts and resulting craters, were mostly confined to its first billion years or so. We'd enjoyed a fairly quiet neighborhood for well over three billion years.[23]

But Apollo's scientific missions brought disappointments as well. Joe Weber, a physicist from the University of Maryland, had successfully lobbied for his special "gravimeter" to be carried to the Moon. Employing a sort of tuning fork that the cosmos could ring, Weber hoped to confirm a fanciful prediction from Albert Einstein: the existence of gravity waves, tiny ripples that move through the very fabric of space and time. The Moon would offer an unusually quiet place to sense these subtle effects, what with no rumbling trucks or portable boom boxes vibrating everything in sight. NASA officials were both excited by and proud of this big-idea type of experiment. A top NASA scientist had even claimed that "the practical utilization of *gravitational* waves may lead to benefits that far exceed those gained from the practical utilization of *electromagnetic* waves." (Given that all radio, wireless, and optical communications use the latter, this is a very tall order.) Placed in 1972 as one of the last Apollo artifacts on the Moon, the gravimeter perked up and started reporting measurements to Earth. But Weber grew increasingly concerned with what looked like a faulty signal. With a heart so heavy one could call it crushed, Weber and his team finally realized that the earth-side manufacturer had carefully calibrated the gravimeter to balance in *Earth's* gravity instead of the Moon's. The discovery of gravity waves would have to wait another half-century. (Weber's method never quite struck pay dirt on Earth either. More recently, gravitational waves were apparently confirmed using, in essence, the world's longest and most precisely monitored rulers, which shrink and grow infinitesimally when the waves move past.)[24]

In the final Apollo mission, scientists sent half a dozen pocket mice along for the ride. With radiation detectors embedded in their bodies, they lived in a hermetically sealed experiment and became some of the only non-human mammals to orbit the Moon and return to Earth. For what it's worth, the five male mice all died, and the lone female survived, reinforcing what medical personnel had found in the early days of NASA: females were at no deficit when it came to space fitness, and their constitutions may even provide advantages for long-term space exploration. Shielding future astronauts (both men and women) is still an unresolved topic for long-duration space travel. There is currently no way to send a person to Mars— let alone to Mars and back—within the allowed thresholds for human radiation exposure.[25]

The Apollo 17 astronauts left behind perhaps the final boot prints on the Moon, and they returned with a rare perspective that may be the most lasting impact of the Apollo program: the now ubiquitous "Blue Marble" photograph. It speaks to our home as an unlikely and precious perch, making things like national borders and political hostilities, at least briefly, appear ridiculous by comparison. They snapped this photograph on the outward journey. Apollo 17 was one of the few missions that brought a spacecraft perfectly between the sun and Earth so that, in looking back, astronauts could see their home fully illuminated. As the mission flew in December, the southern hemisphere tilted toward the sun, and the photo handsomely profiles Africa amid swirling clouds and azure waters. Astronauts had checklists and tasks through the early hours of a mission, and any photo op had to be brief. But none of the three astronauts could resist a look back toward the perfectly lit Earth.[26]

At the end of 1972, with the close of Apollo, the space program itself exhaled, in terms of both funding and personnel. In what some called the "brain bust," the nation had one hundred thousand unemployed engineers in aeronautics and astronautics. While the number sounds absurd, it fits well with the official NASA employment numbers (including both federal employees and those from the many contracting companies), which had dropped by one-third from their 1965 peak of four hundred thousand. Marlowe Cassetti recalled visiting the company that built the command and service modules. "I went into office areas that I recall during Apollo were wall-to-wall white shirts and desks," he says. "And then it was like an empty building, like a morgue."[27] Even these shifts in employment had an abbreviation at NASA. "After the peak work of Apollo," Huntsville's Robert Austin recalls, "it seemed like every six months or every year there was an RIF [reduction in force]. It was really demoralizing. . . . You wondered if you'd escape the next one."

Huntsville had watched von Braun leave suddenly in early 1970, accepting a reassignment to NASA headquarters in Washington, D.C. Many engineers recall von Braun's transition as an ouster, an unfair slap not befitting what he had done for the nation's space program. To his faithful troops, it felt as though NASA leadership was saying, *So long, and thanks for all the rockets.* But von Braun had privately welcomed a change, and he went to Washington for what he saw as an influential planning role. He would help NASA chart its next steps and win over a skeptical political field. Maria von Braun also welcomed the change, as she sought a more cosmopolitan existence. The von Brauns had already sent their daughters, Iris and Margrit, to elite boarding schools in Washington, D.C., and Atlanta.

The nation's cynical political machinery quickly crushed the rocket pioneer's optimism. NASA headquarters found him to be far too impractical for his role, and despite his powers of enthused rhetoric, he could have a surprisingly tin ear for politics. In one congressional meeting, von Braun argued for continuation of Apollo missions by emphasizing how disappointed the astronauts-in-training would be, only to have an angry legislator bark that disappointment was nothing when pilots died every day in Vietnam. As von Braun told an old Huntsville colleague, "I've found out up here I'm just another guy with a funny accent." After watching NASA overlook his recommendations and protestations concerning the shuttle—it was clearly going to be much more expensive than advertised—von Braun retired from NASA (and his lifelong work for two governments) in May of 1972.[28]

Engineers in Houston also experienced Apollo postpartum effects. Mission planner Hal Beck schemed in after-hours sessions with the geologist astronaut, Harrison Schmitt. They examined possible future trips around the solar system. "We had a little renegade group," Beck said, "running a lot of trajectory type stuff that wasn't the proper thing to do. And then he [Schmitt] was taking that stuff to Washington to try to get approval. Then he got kicked in the pants." Washington's appetite for adventure was more than sated. "There was a real long period of bad morale in the program," Beck says. "We just floundered."[29]

Engineer Wesley Ratcliff recalls his decision. "I was not going to leave before we landed on the Moon," he said. But shortly after Apollo 11, "I looked at NASA, how it was slowing down. I liked what I've done, but what could I do for society? I took a pay cut and went to teach." Just as he'd been inspired by his teacher who'd explained the details of Sputnik, Ratcliff, a pioneering African American engineer at NASA, wanted to inspire a younger generation at historically black colleges.[30]

Though many of the retired engineers will cite a lack of vision from NASA's leadership, the nation had simply lost its interest in space. Perhaps America wanted to turn away from that boundary-breaking spirit of the sixties—all of it, from hippies, military strategists, and spacemen alike. Within a year of the Moon landings, more than half of Americans polled said it was not worth the money. And that attitude persisted throughout the 1970s. In 1979, the majority polled still shook their heads: it was cool and all, but nope, not worth it. Apollo, not adjusted for inflation, had cost around $24 billion to achieve the landings. A newspaper columnist noted that developing just the lunar rover, at $38 million, greatly surpassed the entire U.S. automobile safety budget.[31]

The last missions continued the pattern of declining television ratings and interest. The nation suffered images of dead student protestors at Kent State University just two weeks after Apollo 13. Domestic troubles overwhelmed whatever curiosity Americans had left for the barren, gray Moon. NASA tried to take a more savvy public relations angle for the last two missions, armed as they were with better television equipment for the lunar explorations. Planners worked to achieve landings and astronaut excursions that would occur during North American prime time. And they scheduled a special nighttime launch for the final Apollo mission. But the primetime strategy fared poorly, as networks generally opted to keep their most popular programming in place. They then ran a few Apollo clips during the evening news, like so many sports highlights. Apollo 17's exciting night launch crept up to just thirty seconds before a perfect liftoff, at about 9:30 p.m. eastern time, when a technical glitch halted the countdown. The three-hour delay, short for any Apollo engineer, moved past the bedtimes of many interested kids and their tired parents. The networks fumed and tried to fill airtime, not knowing exactly when the launch would come. Those who stuck with it took in an incredible event: an enormous, inverted candle roaring, rising, quieting, and disappearing among the stars.[32]

American culture had turned away from bold optimism to a more negative and fearful stance. The Pew Research Center found that Americans' trust in their government had plummeted between 1965 and 1972, from about 75 percent saying they trusted it "most of the time" to about 50 percent saying that. (And this predated the Watergate scandal.) The mistakes of Vietnam, repeatedly sold and re-marketed to the public, played a leading role. Movies of the era leaned toward demonic possession, societal crises, and dystopian thrillers. Where the first Apollo milestones coincided with the somewhat optimistic *2001: A Space Odyssey*, the later missions played out alongside bleak futuristic fare like *THX 1138*, with a technological society gone horribly wrong, and *The Andromeda Strain*, where a NASA satellite returned a horrible space disease to Earth. A year after Apollo's close, audiences winced as Charlton Heston discovered that his warped, technocratic government was making a food product out of downtrodden citizens in *Soylent Green*.

The Soviets also endured injuries to their space-flight ambitions. Backing away from a manned lunar mission, the U.S.S.R. focused on near-Earth, long-duration work, prioritizing science in space. In 1971, they launched the first true space station, the Salyut 1. That summer, three cosmonauts became Russian household celebrities. They smiled through nightly broadcasts of their work in orbit. After twenty-four days on-board, the crew returned to Earth in a capsule. But as they descended, an air valve opened too early and let precious air slip rapidly to space. The cosmonauts had a manual crank to attempt closing it, but the process was too slow. Within sixty seconds, the capsule was essentially a vacuum, holding three unconscious cosmonauts. Recovery teams, having no idea of the trouble, were shocked to find lifeless bodies in the capsule, and Russia mourned these losses intensely. Some have compared the public outpouring to America's loss of John F. Kennedy. The space station, with no occupants, slowly lost its hold on orbit and burned up in Earth's atmosphere months later. With tensions between the superpowers continuing to thaw, an American astronaut flew to Moscow and acted as one of the pallbearers in the very public triple funeral.[33]

In NASA's painful internal conversations of "What's next?" one immediate option was a collaborative spaceflight between the superpowers. An "Apollo Soyuz" mission proposed for an American craft and a Soviet craft to dock in orbit (a gesture perhaps more symbolic than useful). Max Faget's right-hand man, Caldwell Johnson, had been part of the engineering team to fly into the heart of the U.S.S.R. Johnson recalled a shaky start. "There were great big billboards, all in Russian, showing Uncle Sam with blood running out of his mouth and little babies, tearing them apart and stomping on them, you know. I thought, 'Oh Jesus.'" But the Americans were met with kind hospitality and an eager space team. "The Russian team was first-rate," Johnson said. The difference Americans encountered was mainly one of staffing. "They didn't have much back-up," Johnson said. "For every position you filled, we could have put ten guys there. They just had one or two. But they were really good."

He also encountered predictable levels of Cold War spy action in the periphery of their work. The CIA and army intelligence officers carefully and gravely prepped the engineers, warning them against compromising themselves at any moment: Don't accept offers of vodka, and don't fall for sexual advances. "That was a bunch of crap," Johnson said. "Nothing like that went on. . . . But I could see their cloak-and-dagger guys and our dagger guys hang[ing] around in the background, you know. . . . And they all wore trench coats, just like in the damned movies." Another engineer recalls the KGB tail, friendly but also determined enough to follow the young engineers for a jog around Moscow. In their track shoes, two Americans started their run, and the trench-coat-wearing KGB agents jogged after them, huffing but keeping the engineers in sight.

The Apollo-Soyuz program created opportunities for some at NASA. "Nobody wanted to work it," engineer Cynthia Wells recalled. "They gave me the electrical power system, the environmental control system, the docking module." She and other engineers had to create a special connector between the superpowers in space, because the Soviets kept their ship at a full atmospheric pressure, where the Apollo capsule stayed at just about one-third of a normal atmosphere (but with 100 percent oxygen). The docking module acted like a lock in a canal system, letting people enter from one side, and then slowly having the pressure change until they could move into the other ship.[34]

But the larger question of "What next?" remained a bewildering one for NASA, and the agency never found a unifying answer. How exactly would they follow such an enormous milestone, an engineering masterpiece? There were active plans for "Apollo applications," using some of the existing equipment to launch a type of orbiting station called "Skylab." And Faget's shuttle was in active planning by 1972. But what else? The mission planning folks felt the weight of this question acutely.

"I recall being in a meeting," said Marlowe Cassetti. "I'm embarrassed to say this. It was like we got a bunch of people in a room and our boss got up at the blackboard, said, 'Okay, let's figure out what we're going to do after Apollo. What are some great things we can do? Okay, find a cure for cancer.' It was like, wait a minute." He laughed, thinking back on it. "What madhouse am I in here?" On their chalkboard, just under "cure for cancer," they listed things like a man-to-Mars mission. "We still had a bunch of fairly young people who felt like they could conquer anything, given enough resources." In fact, Cassetti and his colleagues started working a Mars trip. "We did a lot of studies," he said. "Running some trajectories out there, the time of flight, and how long do you have to store the propellants and all that kind of stuff." But the taxpayer's tolerance for expensive adventure had evaporated. A bewildering game of musical chairs ensued at NASA, with all players nervously walking around at once. "Everybody wanted to work on the shuttle," Cassetti says now, but as a favor to his boss, he begrudgingly agreed to take on mission planning for the Skylab program.

Society's eyes were focused inward by this point. Gone were the sleek "space age" designs for cars and kitchens, with gleaming metals and bold colors. The new decade wanted "earth tones," with kitchen tiles that resembled the colors of a rotting forest. Even cars, boxier than before, needed a more "natural" look, with imitation wood paneling.

Would younger Cassetti have expected us, after Apollo, to have shot for Mars within the next half-century? "You know, looking back, I'm really amazed that we haven't done anything like that."[35]

14

1981—FARTHER ALONG

As America approached the first launch in a new phase of space travel, Max Faget, America's original spaceship architect, had decided to retire from NASA. The space shuttle had completed its journey from his gliding, balsa-wood dream to the agency's chief project.

But Faget was disappointed as well. He'd wanted something smaller and more agile, a craft that would be much easier to launch and land. At one early juncture, he'd said he wanted a shuttle to take "two or three thousand pounds of cargo" into orbit. But the political landscape of the late 1960s was long gone. NASA had to win support from a *client*. "And the air force said, 'Well, we'd be glad to use it, but you've got to make the cargo bay much larger," Faget recalled. "They said it ought to carry a payload sixty feet long and at least fifteen feet in diameter." The air force wanted the shuttle to carry its latest spy satellites, and NASA complied.

Faget suffered a life-long itch to nitpick and improve a design. And if he had a central complaint about the space shuttle program, it was a lack of tweaking. A delicate budget dance meant there had been enough money for the initial design and then operating it but little for ongoing modifications. "It's the only case that I know of," he said later, "with no process of evolution to improve that vehicle."[1] He showed the same tenacity for any machine, down to his homemade sailboats. His son, Guy, recalled once taking an elderly Max for a private flight, after Guy had earned his pilot's license. "And I said, 'What do you think of this?' It was a fairly nice plane." Max focused his full attention, circling the parked craft and finding puzzling details. "This is unbelievable," he said. "Why do they have all of these things? I can't believe they couldn't make this plane faster." In the end, Guy says this is how he will always remember his father, analyzing and fixing.[2] In his post-NASA years, Faget helped found Space Industries, seeking to entice private capital upward, into orbit or beyond. Though a recipient of NASA's highest honors and a legend within the agency, he never achieved much of a national profile, from the parades of the early 1960s until his death in 2004. And, by all accounts, that suited him just fine.

By 1981, Henry Pohl was on his way to taking over Faget's role, as NASA's director of engineering in Houston. Pohl had stayed true to his first love from the 1950s: the beguiling power of a rocket engine. As chief of propulsion and power at the Johnson Space Center, he oversaw the many technical challenges of the shuttle's main engines, a cluster of three hydrogen burners directly descended from those in the Saturn V. (The final design of the main shuttle engines was robust enough that they are being dusted off and refurbished for use in the nation's twenty-first century efforts.)

The year 1981 felt like a reawakening for NASA, preparing for a return to manned spaceflight. The shuttle program had survived brushes with cancellation, and the agency was back on its feet after a difficult decade.

After the last Moon mission in 1972, Capitol Hill had excoriated NASA for its employment record. Scanning the ranks of white, male engineers, Congress openly wondered if the agency had just ignored the equal opportunity mandates of the 1960s. While the federal government had increased minority employment to 20 percent of its workforce by 1973, NASA's sat at 5 percent. By early 1974, under scrutiny from multiple congressional committees, NASA found itself under legislative oversight. The agency's days as a government darling were officially over. Longtime NASA man George Low, who had boldly pitched the 1968 Apollo 8 flight around the Moon, had become a senior NASA administrator. Stewing in helpless 1970s frustration, he made an incredibly telling remark. It wasn't just that equal opportunity and affirmative action at the agency were "a sham," he testified, but more that NASA's engineering culture possessed "total insensitivity to human rights *and human beings*" (emphasis added). For NASA managers, given the ambition of their lunar task, they'd perceived no time to worry about the finer points of a diverse workforce. To some, even considering their employees' basic humanity felt like an abstract luxury.

Charged with investigating NASA's hiring record in the early 1970s, Ruth Bates Harris later recalled a poignant meeting with Wernher von Braun in Washington, D.C., after he'd left his longtime post in Huntsville. Harris had read a newspaper article describing Germany's underground V-2 rocket factory, *Mittelwerk*, with its inhumane slave labor. She went to von Braun's office and gently but firmly asked if the article was accurate. With a heavy heart, he acknowledged the factory but said the article distorted the story. She told him she had to let the top brass know of these troubling parts of his past. He nodded. She said he didn't fight with her at all, but was instead "sad and understanding." In reviewing the work of all centers trying to follow affirmative action guidelines, her eventual report said von Braun's time at Marshall had shown the most "courage and conviction" in pursuing minority hiring, even if making little progress.[3]

The rocket pioneer's earliest work, long ignored in the United States, started bubbling to the surface. Groups of protestors often plagued his public appearances, drawing attention to his Nazi past, and even the amiable television host Dick Cavett surprised von Braun with difficult questions about the V-2 weapon. The pioneer's tarnishing reputation eventually led President Gerald Ford's advisors to deny von Braun a Presidential Medal of Freedom in 1976.[4]

The Marshall center von Braun left behind had changed quickly after Apollo. As its third director, following von Braun's longtime collaborator Eberhard Rees, NASA appointed Rocco Petrone. An engineer and army officer who had once worked at the Redstone Arsenal in Huntsville, Petrone had directed launch operations at the Kennedy Space Center during Apollo. The engineers sensed the new leadership pushing out a lot of the older Germans by encouraging or even forcing early retirements. Many younger German engineers found themselves transferred to other centers. Some of the retired Huntsville engineers still speak of this brief era—from early 1973 to spring of 1974—as the "Petrone Massacre." Inside NASA headquarters and at other centers, some had long referred to the Marshall center in snide tones as "*Huns*ville." Many engineers perceived headquarters intentionally breaking up the group of Germans that America had once forcibly gathered.

Arguably as responsible as any other American citizen for the successful Moon missions, Wernher von Braun passed away at age sixty-five in 1977. After Apollo, he had battled cancerous tumors, losing a kidney to surgery in 1973 and part of his colon in 1975. Until the end, he maintained his enthusiasm and as busy a schedule as he could manage, though with increasingly less of his trademark energy. In his last essay, "Responsible Scientific Investigation and Application," he aimed his vision at earthly matters. He faced, in his own words, "the grim problems besetting humanity" with a mix of pragmatism and philosophy.[i]

As von Braun spent his last painful days surrounded by friends and family, the latest federal budget ordered NASA to stop monitoring all the remote experiments left on the Moon. Though the instruments would continue broadcasting temperatures, moonquakes, and other data for years to come, Earth simply quit listening in 1977.[5]

As the space shuttle rose from Earth in 1981, so did Apollo's popularity. By 1989, three in four Americans said Apollo money was well spent. Perhaps the shuttle, the most complicated device ever constructed, served as a reminder of space flight's extreme challenges. In an earlier technological age, with slide rules, horn-rimmed glasses, and grainy video, we had *gone to the Moon*.[6]

Apollo had triumphed before my generation was really conscious. But we had steady reminders. Kids binge-watched a new phenomenon in 1981: Music Television. An Apollo astronaut on the Moon provided its flagship symbol. Where the stars and

---

[i] After his death, von Braun's American star fell still further. Within a decade, not only did his Nazi SS membership finally come to public light, but so too did his collaboration with *Mittelwerk*, where thousands of slave laborers perished in hellish conditions.

stripes should have been, a glowing "MTV" insignia buzzed on the flag. "We had this idea of copying the biggest TV event in world history," the network's creative director later explained. "The man walking on the moon."[7]

Growing up in the shadow of the Johnson Space Center, most of us kids took for granted that the nation's space investment had reaped incredible rewards: *For every dollar spent on NASA, America gets five dollars back in technology.* We recited this like a prayer, even though, given such an incredible claim, the examples should have been more obvious and plentiful. People brought up the orange-like drink Tang and the grip of Velcro fasteners, but even those light-hearted examples had predated the birth of NASA.

The impact of the space age certainly goes far beyond a few indentations in lunar soil, but a careful analysis questions the exact return on investment. Scholars trying to untangle the threads connecting the space age to major technologies find two indisputable examples. First, Earth-orbiting satellites have had a deep and pervasive impact on commerce and communications. Without our rush to meet the challenge of Sputnik, we would have been years or decades slower in putting "comsats" in orbit around Earth. They've revolutionized the connectivity of the world, and their cousins, the weather satellites, also deserve mention. Having a much better understanding of what storms are headed where, and when they'll arrive, has saved lives and boosted the efficiency of global shipping.

The next most important influence of NASA is less obvious: It greatly upgraded food safety for a huge swath of humanity. My father has always explained NASA's approach to spacefaring as "fo, fo, fuzz"—a goofy pseudo-acronym for "fail operational, fail operational, fail safe." For a mission flying through a dangerous vacuum and coming back through the atmosphere, engineers designed each system so that it could have one failure, and still keep going. Then, it could suffer a *second* failure, and the mission would soldier onward (its second "fail operational"). An unlikely *third* failure within the same system might cut the mission short, but it should not endanger the crew ("fail safe").

The agency took this "fo, fo, fuzz" mindset to astronaut nourishment just like it would for a fuel line. NASA had asked the food industry as early as 1959 how they ensured food safety and were surprised to learn that they really did nothing of the kind. The industry just responded to outbreaks of foodborne illness after the fact, with no rigorous method to prevent them. Medical personnel and mission planners didn't like the idea of responding to food poisoning in a small, orbiting capsule. Instead, the agency developed a very NASA-sounding system, the "Hazard Analysis and Critical Control Point" approach to protecting astronaut meals. Working with an army laboratory, NASA set a maximum bacterial count per gram of food, a radical idea for the early 1960s. The agency developed a series of seventeen "control points" in food assembly like they would for any space mission. Stop at this step, they told the food industry, and answer a few key questions. Then make a "go" or "no go" decision for, say, a tin of deviled ham. If it moves forward to the next step, you will again stop to review the data and make another informed decision. NASA unveiled this more broadly to the food industry in 1971, and by 1981 the sensible protocol infused the seafood preparation and canning industries. Eventually, an array of international food producers adopted the engineering-informed safety system, and it now protects a sizable fraction of the world's food supply.[8]

Beyond these two giant NASA examples, the true impacts become more muddied. Some sources credit the space program for direct impact on medical technology. Engineers had worked with doctors to track an astronaut's vital signs during these stressful space missions and for days afterward. This required new devices, more miniature and reliable than anything available in a hospital, that could send live updates from afar. Larry Bell described the biological containment protocol for re-turning astronauts being adopted by the U.S. Centers for Disease Control, and he relayed a direct migration of NASA's astronaut biosensors to the operating room. Bell's group brought some of their sensors to Dr. Denton Cooley, a pioneering heart surgeon at the Baylor College of Medicine. It gave Cooley's group their first access to round-the-clock electrocardiogram (EKG) monitoring for a patient and greatly shrank the equipment required. But that wasn't what most impressed the surgeon. "It was the first time Dr. Cooley had seen that clean an EKG," Bell recalled. The doctor had asked where all the extraneous wiggles had gone. The NASA engineers, as usual, had squeezed every bit of noise that they could out of any measurement and, in this case, found a heart's pristine underlying beat.[9]

And how about computers? In January 1981, the Apple Computer company held its very first shareholder meeting, after their initial public offering had attracted more capital than any other since the Ford Motor Company's in 1956. By August of 1981, International Business Machines entered the new "personal computer" market. Can we say, as many do, that NASA's early embrace of semiconductor chips hastened America's eventual immersion in and leadership for the digital age? Didn't Apollo push computers to become smaller, more lightweight, more robust, and energy effi-cient? The story is not so cut-and-dried.

Jack Garman, one of NASA's chief computational engineers, later said, "The Apollo guidance computer was no PC." But he also believed that, aside from making a computer smaller, Apollo required a type of interactivity that computers had rarely offered before.[10] The Apollo guidance computer had to react, in real time, to changing conditions, and it had a relatively novel "interface" where astronauts could ask questions and request maneuvers minute to minute using a keypad. Before Apollo's version, most computers were hulking machines that worked like oracles.

You left an offering of punched cards and returned in the morning for the computed wisdom.

However, historians of technology note that NASA, in its understandable rush for the Moon, opted wherever possible for proven circuit technologies instead of pushing for new innovations. And while NASA placed some of the first big-ticket orders for semiconductor chips, the military, ever the valued tech customer, was buying the rest. The military-industrial complex would have kept the baby computer industry crawling forward, with or without NASA. Since the military wanted computers to fly in guided missiles, for example, they were applying significant pressure for computer miniaturization and reliability. The dawn of the personal computer, sans Apollo, would probably have still lit our homes with ghostly blue light in the 1980s and 1990s.[11]

One of the earliest adopters of the personal computer found himself finally benefiting from what once seemed a useless skill. "I took typing in high school at my dad's insistence," Marlowe Cassetti recalled. "He deemed I wasn't smart enough to get into college." (Dr. Cassetti had hoped that if Marlowe was drafted into the Korean War, typing skills might keep him far from the front lines.) In 1974, an electronics magazine article had caught Marlowe's eye: how to build your own microprocessor. By the end of that summer, he had an 8-bit homemade computer up and running, and he founded one of the first home computer clubs in the Houston area.

After years of laboring through stacks of punch cards, he'd never considered himself a computer enthusiast, but his work life shifted in a similar, digital direction. After completing work on the Skylab orbiting space station in 1974, Cassetti had started a new job working with the eventual shuttle's computer software. "On Apollo, Gemini, Skylab, and everything, there was nothing that the crew couldn't do if the computer went out," he said. "But with the shuttle, from the start it was a fly-by-wire system, as they call it." In one sense, the shuttle was more like the Saturn V, making many adjustments moment-to-moment via an automated brain, with little to no human input.[12]

As the shuttles began achieving orbit in the early 1980s, Cassetti, like Faget, decided it was time to retire from NASA. Making a last drive through the main gate of the Johnson Space Center, he would have passed an unmistakable reminder of Apollo.

The canceled Apollo missions, 18, 19, and 20, provided NASA with enormous leftovers. By 1981, the Johnson Space Center had set out one of the three remaining Saturn V rockets for all to see. Built to be vertical, this rocket reclined stage by separate stage on the earth. The black-and-white paint matched all the nearby buildings of NASA, white and black by layers. For a twelve-year-old kid, they were something from a different era, like a television showing reruns of *Leave It to Beaver*. It seemed like another relic of America's wholesome, bygone days.

I remember going to the rocket with my father. He told me it was more powerful than the shuttle's launch system and that America couldn't even build a Saturn V anymore. We would barely know where to start, he said, and some of the blueprints had been lost. (That was true for a while. The blueprints wandered a bit, but NASA could account for nearly all of them as of 1987. Some of the specialized tools required to build the parts, however, had already been sold as scrap metal.)[13]

Eventually, he wondered in our evening talks if federal agencies should come with built-in expiration dates. So, NASA would have ended? Well, he said, a new agency should have been born with a new, clear set of space exploration goals. To his mind, and those of many other Apollo engineers, an incredible, do-anything organization had succumbed to overgrown, unfocused bureaucracy.

My father had become a branch chief in the later Apollo years and smoothly transitioned into a "mission analysis" role, helping develop a ubiquitous piece of planning software. Poignant in hindsight, he also ran a "failure analysis" project in the early shuttle days, following a NASA tradition of planning for the worst outcome and understanding, as best they could, the odds of survival and success. His team identified some weak points—including the solid rocket boosters—but he recalls powerful voices dismissing some of the predicted odds as "Brown's horror stories." As always, the agency had to weigh worries and warnings against payoffs and progress.

What I remember most, from the shuttle planning era, is his morning routine. His excitement and his sled-dog-like energy for each day still embody, for me, these NASA engineers. As he buttoned a short-sleeved shirt and fixed his necktie, he belted out parts of his favorite hymn: ". . . Farther along, we'll understand why. Cheer *up* my bro-*ther*. . . ." Once he had clipped his NASA ID badge to his shirt pocket he'd start pacing at the front window. As America roiled in gasoline shortages and carpools, Robert Brown awaited the sound of his coworker's car. Standing at the front window, he stared at an empty street barely lit by dawn. "Come on," he'd say to himself, to his ride, to the day and the work ahead.

15

TODAY—MEMENTOS AND RETURNS

In early 2018, Marlowe Cassetti inverted our normal routine: He sent *me* a question. He wanted to know the working title of this book. After he received my reply—*The Apollo Chronicles: Engineering America's Moon Missions*—he thought it over before adding his two cents.

"I like 'Engineering the *First* Moon Missions.'"

First? Did he mean there will be more to come?

"Yes. . . . Optimistically."

While the surviving Apollo engineers express a great range of predictions, from hopeful to doubtful, the younger voices at NASA have worked out a path for new, Apollo-like missions.

Over the last several years, the old centers have shown new signs of life, in many cases sprouting from Apollo's old roots. At the Cape, the enormous Vertical Assembly Building proudly wears fresh paint, and engineers plan to bring the old crawlers back to life. Von Braun's directions for the "mobile launch" scenario are back in business: stack a massive rocket in the world's most cavernous garage, bolt it to a movable launch platform, and then inch it toward a launch pad.[i] NASA's new "Space Launch System" looks like a newer version of the Saturn V with shuttle-like side-boosters. (While not quite as tall as a Saturn V, the SLS will fight gravity with slightly more power.) The rocket's main engines, literally handed down from surviving space shuttles, reprise the liquid hydrogen burners of Apollo's upper stages. The new crew capsule expands the old command module to include a fourth seat, but it will return to Earth the old-fashioned way, like a meteor just slow enough to survive, holding huddled astronauts with their backs against the heat shield (see Figure 15.1).

Meanwhile, the view from atop the old Saturn V test stand in Huntsville features new skeletal stands rising nearby to help test-fire Mars-hopeful rockets. Many of the old facilities remain at the Marshall Space Flight Center. The Saturn V static test stand still looms like some sort of post-apocalyptic castle, dwarfing in sheer mass

---

[i] The private firm Space X uses the old Apollo launch pads for its busy schedule, but it prefers to prep its rockets at the launch point.

FIGURE 15.1 The Vertical Assembly Building at the Kennedy Space Center, freshly painted and ready for business again. For scale, one stripe from the painted flag could accommodate the width of a standard car. Opening one of the 456-foot-tall doors takes roughly forty-five minutes. Starting from the bottom, each square panel slides upward to nest behind its higher neighbor. (Photograph by author.)

all the other facilities. In its old concrete body, a network of hallways, elevators, and stairways connect rooms alternately holding huge pumps or clustering bales of power cables. On one of the higher platforms, the metal ramparts underfoot show plenty of rust.

In hindsight, it's difficult to believe rocketry's ascent from humble mid-century origins to the Moon. In addition to the 1960s Saturn test stand, the Marshall Spaceflight Center maintains an older one as well, dating to the 1950s shortly after von Braun moved to Huntsville. This one looks like a poor excuse for a nineteenth-century oil derrick: sturdy, hewn by hand, and standing just twice the height of a basketball goal. Von Braun borrowed whatever parts he could find, because any expenditure over $20,000 would have brought additional congressional scrutiny. In the pre-Sputnik days, rocketry was a fanciful pursuit.[ii] The structure stands as a reminder of the space program's humble origins: repurposed timbers the width of human limbs and a flame-bucket the size of a shallow grave mixed here with other-worldly energy, ingenuity, and ambition.

If one drives into Huntsville, Alabama, at night, a massive, uplit rocket—a full-sized replica of the Saturn V masterpiece—dominates the landscape. As the tallest structure in town, it anchors the U.S. Space and Rocket Center, a brainchild of von Braun's. The center opened in 1970, as a gear-headed, kid-friendly technical museum, and it remains Huntsville's primary tourist destination.

Huntsville is still very much Rocket City. A full-thrust-ahead attitude pervades the Marshall Space Flight Center and a lot of the town itself. At the airport, a larger-than-life Wernher von Braun makes intense eye-contact with arriving passengers. His face anchors a large mural summarizing Huntsville's rocket history and projecting hope for the future. The new Space Launch System, a natural descendent of von Braun's work, anchors the mural's other end, seeming to say the intervening years were just a pause, a deep breath.

As of this writing, the nation has a Space Policy Directive signed by President Donald Trump on December 11, 2017. It calls for a return of American astronauts to the Moon, using a public-private partnership. From there, the plan leads to Mars and, eventually, other stops around the solar system. The directive provides no specific timeline or special stream of funding, and NASA currently consumes about 0.5 percent of the federal budget, compared to about 4.5 percent in Apollo's heyday.

There have been several other presidential proclamations following Kennedy's challenge. In 1989, George H. W. Bush honored the twentieth anniversary of the first lunar landing and announced the "Space Exploration Initiative." He declared that America would return to the Moon, establish a base, and then journey to Mars, all according to a logical thirty-year plan. It never got off the ground with Congress or the public. He had believed in what he was saying, but years later he reflected that he'd been "set up" and misled by space advocates.[1] Despite that disappointment, his son George W. Bush unveiled 2004's "Vision for Space Exploration." It also called for a return to the Moon but requested no new federal spending. Time will tell if America will ever again take our species beyond Earth orbit, but the vision of a president single-handedly pushing the nation outward appears misleading or at least irrelevant

[ii] My thanks to NASA historian and tour guide Brian Odom as well as tour companions and retired engineers Tom Parnell and Don Woodruff.

to modern times. Even Apollo emerged less from President Kennedy himself and more from Cold War panic.

If you ask the Apollo engineers about Mars, many hedge their answers. They don't want to be negative, but they admit this effort might make the journey to the Moon look like a cakewalk. Recent studies of long-duration space missions (e.g., at the International Space Station) aren't encouraging. The human body exhibits a great deal of stress away from its natural home. And instead of days, any trip to Mars will take many *months*, leaving astronauts riddled with cosmic rays and risking many more rolls of the solar flare dice. Chances of brain damage and leukemia would sky-rocket. Some scientists estimate, for instance, that about one-third of an astronaut's brain cells would have been struck by cosmic rays for a thirty-month round-trip to Mars. But aside from the perilous transit, at least Mars would appear more habitable than the Moon at first glance, right? The planet has a thin residual atmosphere and at least a warm-looking color. But the soil, far from being crop ready, offers unique dangers. It contains a compound hazardous to the human body at even small doses.[iii] And the planet has no magnetic shield to protect life from the pitter-patter of cosmic rays. Unless Mars pioneers build underground, they would suffer extreme radiation exposure.[2]

Obstacles notwithstanding, one of the most eager groups for a new Apollo-type era lives within planetary science. Apollo's gritty bounty—the scientific measurements and rocks returned from the Moon—is considered the gold standard not just for understanding the Moon but also for calibrating the solar system's history. Scientists crave more material from the Moon and, if possible, other planets. To an extent that would surprise most of the Apollo engineers, scientists are still gleaning discoveries from the old Apollo data and the remaining lunar material. Using modern techniques on old data or old rocks can truly squeeze fresh blood from old, gray turnips.[iv]

Early on, Apollo's advocates had promised an understanding of how the Moon came to be. Today, we still haven't nailed down a unanimous picture of the Moon's origin. The composition of the Moon rocks (similar to Earth's crust) and the prevailing age of the oldest rocks (roughly ten million years younger than Earth by recent estimates)[3] support a "giant impact" hypothesis. Though it's not a closed case by any means, the prevailing narrative unfolds as follows. If we compress the 4.5 billion years of our planet's existence to just forty-five years, with Earth now entering middle

---

[iii] Perchlorate compounds interfere with proper thyroid functioning. They would be ever-present in the treads of any astronaut's boots.

[iv] With many thanks to scientists Caleb Fassett, Debra Needham, and Renee Weber for their time and patience in sharing the world of modern lunar science. Any mistakes in the text are mine alone, in my attempted simplification.

age, then when Earth was still but a babe, an interloping body (or a swarm of them) sideswiped Earth and knocked a bunch of its crust into a nearby orbit. Gravity pulled this material together into the Moon, and gravity likewise helped wounded Earth return to a spherical shape.

Unlike Earth, the Moon's surface preserves its history with museum-like precision, pristine for billions of years.[4] It provides a unique sort of time machine for astronomers. As scientists today return to the Apollo Moon rocks, they can literally see and even sniff history to an unprecedented degree. For example, recent work has probed tiny, sealed bubbles within the Moon rocks, using specialized needles. In letting these bubbles exhale a tiny breath held for billions of years, scientists can smell exactly what the young Moon was cooking in its turbulent youth, including a surprising amount of water vapor.

But Apollo traded with the Moon—as it brought material to Earth, it left mementos behind as well. One of the most significant leave-behinds needed no batteries or transmitters: just a simple set of mirrored reflectors, originally proposed to NASA by a Princeton graduate student. Pinging these reflectors with laser beams allows Earthlings to measure the distance to the Moon with incredible precision. This measurement is still a central one for physicists debating the finer details of gravity. The Earth-Moon distance isn't simply increasing an inch-and-a-half per year as the Moon recedes—it's also constantly fluctuating. The Moon's devilishly complex orbit (elliptical but not without tiny irregularities) provides humanity's best available ruler against which to test, for instance, Einstein's general relativity against other, newer theories.[5]

The later Apollo missions also left active experiments on the Moon, transmitting measurements to Earth that NASA scientists recorded onto spools of magnetic tape until 1977. These precious reels, each about the size of a dinner plate and numbering well over ten thousand, won't be with us forever. The magnetic tape degrades, and even maintaining the antique equipment that can wind and read the tape presents an ever-larger hurdle.

One of the most bountiful data sets wiggled forth from the remote seismometers measuring each shudder in the lunar soil. By combing these old data sets with modern, digital tools, scientists have identified more than four hundred previously undetected quakes that happened deep within the Moon. Researchers continue building a better understanding of the Moon now, year by year, using the old recordings. Scientists have also identified the seismic signatures for scores of meteoroid impacts from these data sets. Since Earth has a thick atmosphere, many of our meteors burn up undetected. But the record of thumps on the Moon tells the straight story, providing astronomers one of the most reliable measures of the solar system's "bombardment rate." That rate, pegged in the Apollo seismograph data and carefully married to a catalogue of craters, allows astronomers to extrapolate backward in time, to glimpse the solar system's more violent childhood when the rate was much higher. Billions of years ago, as Earth and the Moon spun as mere children in the cosmos, they withstood a relative hailstorm compared to today's drizzle.

The Apollo equipment packs, strewn on the Moon like the remains of nerdy picnics, transmitted all sorts of other information as well. Tragically, a lot of it has wandered off and disappeared. NASA, in shedding its Apollo spending habits, couldn't store all the tapes. The agency sent boxes full of them to scientific experts and, presumably, safe and secure storage within their various universities. But buildings flood, professors retire, and university administrators tend to wrinkle their noses at obscure, musty boxes. Approximately *half* of Apollo's remote sensor data is now missing, and the evidence trail for thousands of tapes has grown cold.[6]

In many homes of Apollo engineers, you can find a wall or corner set aside for their own mementos. Key photos of missions and banquets mix with plaques of recognition and, less often, models of rockets or spacecraft or even bits of equipment. As with Mrs. Faget's tearful moments following a 1962 parade, the engineers' spouses are often the ones to express blunt pride about the mountains of overlooked work. No matter how much time you spend with the engineers, you come away feeling you've only skimmed the surface of this history. Many of them mention losing colleagues more and more frequently. Even while scheduling interviews for this book, some have passed away, taking with them unique stories of near misses and late-night "eurekas."

A great many of the Apollo engineers have returned to—if not their origins— initial loves. Marlowe Cassetti, now retired in western Colorado, has rekindled the rapture that began with his first balsa-wood glider. "I probably have a dozen," he says of his model airplane fleet, but he adds that they could use some work, after a few crashes. He helps organize a couple of model airplane clubs. They meet year-round, and Cassetti's Houston-inspired chili is a crowd favorite when they land their planes in fresh high-country snow.

He's watched the groups dwindle in recent years. "Unfortunately, many of the people are older, retired," he says. "And we don't have a lot of young people come in." The younger fliers seem to prefer drones. These come pre-assembled, require less time, less attention to detail—less engineering. Drones don't need long, flat runways cleared of rocks either.[7]

Frank Hughes, one of Apollo's simulation wizards, stays as busy as ever. He works an array of engineering consulting gigs, his active mind seeking good problems like an aardvark tearing into logs for insects. The cold winter of 2018 found him in a west Georgia rock quarry. He served as an advisor for the film *First Man*, which portrayed Neil Armstrong's path to the first lunar landing. Frank witnessed a whole new type of simulation, with the bleak quarry standing for the lunar surface. The cast and crew occasionally halted filming to wait out some very un-lunar snow showers.[8]

Henry Pohl has circled back to cattle ranching. And he has returned to the family land near Ezzell, Texas. Now in his mid-eighties, he owns several hundred head of cattle grazing on about a square mile of warm, green land. Ezzell is as rural as one can still get in Texas, a three-hour drive south of Austin, with two-lane state roads passing through alternating German- or Spanish-settled communities like Schulenburg and La Grange. No cell phone's mapping tools can guide you to Pohl. Gravel county roads eventually give way to a private dirt road that rumbles over cattle guards and past a hand-carved "Welcome to Paradise" sign. He and a few family members live in homes scattered near the original Pohl homestead—the one he wired for electricity in the late 1940s—and it stands ringed with weeds and nostalgia. Most of its last coat of white paint fiercely clings to dried gray boards (see Figure 15.2).

If you've arrived on a warm day, Henry, tall and cordial, might be sporting bare feet. His fierce blue gaze reminds you of his memory's power. When he tells a story, he inserts pauses. He is not searching for a word, but letting the story prepare a pivot,

FIGURE 15.2 Meeting for the first time since their NASA days, Robert Brown (*left*) and Henry Pohl (*right*) reminisce near Ezzell, Texas, in 2017. During the 1940s, Pohl wired the family home in the background for electricity. (Photograph by Dana Smith.)

surprise, or shift. "It was about forty degrees out, and it was raining. I was terribly cold and wet," he says of his first tractor ride. He pauses a beat, before adding, "but I was the happiest boy on this Earth."

Pohl has devoted most of one room to mementos. A large model of a space station sits in one corner. Henry and his colleagues designed it later in his career, but NASA abandoned it, he says, when they decided to partner with the Russians. A thruster manufactured for Apollo sits on a shelf like a bowling trophy. He grabs a piece of the pint-sized Saturn he built for von Braun as well, and he points out the double-walled structure. "In between," says Pohl, "we ran water down this side of it and up the other for cooling."

Pohl likes to take guests around his ranch land. At one stop, near the old homestead, Henry points out a structure he built as a boy. He and his brother cut the timber one year, let it dry, and then erected a barn the next. "It looks better than it functioned, that one." Henry failed to predict the width of cars and farm equipment as the twentieth century roared past. But this structure from the mind of an untrained boy, an engineer-to-be, is remarkable.[9]

My father's mementos, kept in a box, feature a now-familiar collection: mission pins and patches, letters of thanks and commendation from NASA leaders, his early papers plotting Moon routes, some mission photographs, and a bit of shuttle tile wrinkling within a Lucite block. As of this writing, my father is as restless as ever if not more so. He's spent years volunteering for veterans' organizations and, with my Mom, for before-school reading programs in elementary schools. The volunteer opportunities start to dry up for people in their eighties though.

At my parents' central-Texas retirement home, he is known as a kidder. So, when people ask what he did for a living, I suspect some don't initially believe him when he flatly replies "rocket scientist" with no elaboration. In the spring of 2018, he gave a talk about the Apollo years, coordinated to follow a screening of the movie *Apollo 13*. About fifty curious retirees gathered to hear about those Moon missions. They sat attentively as he went through some slides and stories, and then some hands went up. People seemed very appreciative, even impressed. But most of their questions centered on the astronauts and the minute-to-minute mission drama, as opposed to the painstaking wiring, test runs, and computing—all the late-night homework, sweat, and worrying—that lifted Apollo to success.

16

HOW WE DID IT

It was like we was both back in older times and I was on horseback goin through the mountains of a night. Goin through this pass in the mountains. It was cold and there was snow on the ground and he just rode past me and kept on goin. Never said nothin. He just rode past and he had this blanket wrapped around him and he had his head down and when he rode past I seen he was carryin fire in a horn the way people used to do and I could see the horn from the light inside of it. About the color of the moon. And in the dream I knew that he was goin on ahead and that he was fixin to make a fire somewhere out there in all that dark and all that cold and I knew that whenever I got there he would be there. And then I woke up.
—Sheriff Bell describes a dream of his father, in Cormac McCarthy's *No Country for Old Men*

A group of NASA thought leaders, including many of the original group from Langley along with a few astronauts and key administrators, gathered a couple of years after the Moon program's conclusion. "Twenty-five or thirty people that had all worked on Apollo," engineer, flight director (and eventual director of the Johnson Space Center) Gerry Griffin recalled, "We went out to Caltech and spent about three days contemplating our navels.... what we had done, why we had done it, how did we do it."

"We talked a lot of technical stuff, some politics," Griffin said. And he tells a story of attendee Neil Armstrong, the first man to set boot on the Moon. "He didn't talk a lot, but when he spoke, people listened carefully." Armstrong was in many ways not a typical test-pilot-turned-astronaut. While known for being quiet, he was never what someone would call shrinking or invisible. His thoughtful presence carried significant weight in a room. He'd been a bookish kid and grown to become an oddly earnest high schooler, studying calculus on his own time and carefully assembling a wind tunnel in his parents' basement. Instead of turning wrenches on hot rods, he learned to fly before he could drive. "I am, and ever will be," he once said, "a white-socks, pocket-protector, nerdy engineer."[1]

After everyone else had finished speaking at the Caltech gathering, Armstrong calmly rose and went to a chalkboard. He drew four bell-type curves, spaced slightly apart. "They looked kind of like mountain peaks," Griffin recalls. The astronaut labeled the curves: Leadership, Threat, Economy, and Talent. And he said to the room, "My thought is, when you get all these lined up, you can't stop something really big from happening." A bold (and in some ways, desperate) president. The threat of the Soviet Union. A roaring economy with budget surplus. And a newly educated bunch of young adults. When the curves aligned—a rare enough event in history—Armstrong suggested an Apollo could bloom. "We were trying to figure out what allowed us to get this done," Griffin recalls. "He kind of summed it up in a way that everyone said, 'Of course—you're exactly right.'"[2]

The analysis of aligned curves can help explain why we haven't quite restarted a program of sending humans to explore the cosmos, despite many presidential initiatives. But the four factors cannot account for four hundred thousand voices and their feat: joining forces in peacetime on a breakneck, ambitious project. Looking at our species and our collective history, Apollo is an outlier. We are probably wired to comprehend and associate with only so many people. Many researchers believe that our human specifications, as tuned by evolution, have us naturally understanding and organizing groups of about 150 humans.[i] Any greater number stretches a brain uncomfortably. (Some see the rising angst of the internet age—with its disconcerting swarm of voices and faces—as a predictable result.) Since the Apollo engineers were part of an anomaly, we can ask them what they saw and experienced as the glue.

First and most obviously, Apollo had a goal of perfect clarity: As advised by Wernher von Braun and Bob Gilruth, President Kennedy and Vice President Johnson stated the mission criteria in a way that every engineer, every technician, and really everyone on the planet could understand. In 1961, Wernher von Braun said Kennedy's announcement "puts the program into focus. . . . Everyone knows what the Moon is, what this decade is, what it means to get some people there."[3] The only gray areas were left on the Moon itself. And such a clear goal underwrites many of history's rare stories of dramatic success. Ironically, the clarity of Apollo made the next steps difficult at best. How could NASA get that many people on the same page again?

Getting to what worked and the answer to how they did it, I'm going to group the answers together into four wide buckets. They appear here in the order of how often these factors come up in interviews with engineers.

AGENCY IN THE AGENCY

The first time I asked my father "what worked," he immediately spoke of the way old NASA doled out big scoops of responsibility. Nearly every Apollo-era engineer

---

[i] This is "Dunbar's number," named for anthropologist and evolutionary psychologist Robin Dunbar. Some now refer to the idea as "the monkey sphere," given the millions of years that fine-tuned the wiring of social brains.

will tell you they *owned* their work. Just return to July 1969 and hear each of them hyperventilating; their one part of one device or calculation could be the one to ruin the enterprise. Whether it was miscalculating the fuel supply, worrying about a breaking ladder, missing a tiny flaw in the lunar lander's ascent engine, mistyping or mis-weaving the computer code, or mistaking the Moon's spastic gravity, the engineers lived, breathed, and sweated their contributions.

The responsibility bestowed and respect given to each employee was woven into the organization's DNA before it was NASA. The twin cultures of the Langley research laboratories and von Braun's missile team in Huntsville each worked in this way: loose leashes and high expectations.

Max Faget's best buddy drove with him to Langley when both were just scruffy young war veterans. "It was the greatest place to work in the world," Guy Thibodaux later said. "It was what you'd call a bottoms-up outfit. All the ideas started at the bottom and came up to the top, and the managers had enough sense to kind of steer you or guide you, but never interfere with your work. . . . Just 180 degrees from the way things are right now [in 1999]."[4]

The engineers routinely mention the trust. If your part, or your system, or your simulation was supposed to work, everyone around you assumed it would work by the required deadline.[ii] It *had* to be perfect, because for many years the project didn't impose (or offer) extra levels of double-checking.

"It started at the top," says Gerry Griffin. "Nobody at NASA was micro-managed by their boss. From the beginning . . . the lowest rank could speak up. The leadership wanted everyone to be heard. They let us do our thing. They . . . empowered us and held us accountable." He cites the lightning-struck Apollo mission as an example. His bosses trusted him to assess all the spacecraft data from his engineers and make the decision: onward to the Moon or come home? His supervisors allowed the flight director to do his job. Similarly, in the wounded Apollo 13 mission, he recalls telling NASA leaders that they would go behind the Moon and sling their way home instead of turning the broken ship around. The brass listened and then asked simply, "What do you need? What can we do to help you succeed?"[5] In the 1960s, this sort of reasoning dominated Gilruth's outfit in Houston and von Braun's in Huntsville.

A striking example of trust—of responsibility overcoming rank—sits tucked away in a devilishly tricky task: How would NASA scoop those conical capsules from the

---

[ii] Before painting too simple a picture, we shouldn't forget that engineers who didn't look like everyone else sometimes received smaller allotments of trust or had to earn it one task at a time.

ocean, without them turning over and sinking, without them pulling a helicopter into the ocean, and without them smacking against the side of a navy destroyer? Engineer Peter Armitage recalled an early test that was not going well—a dummy capsule was about to sink. As the engineer in charge of the test, he stood with his boss watching a capsule listing—they knew they had to intervene quickly. "Do you want me to swim a line out?" his boss asked. Armitage laughed at the memory of this. "I said, 'Sure.' I mean, he's asking *me*, because I'm the test engineer and he recognizes who he's put in charge." His boss stripped down to his underwear, swam to the capsule and attached a rope.[6]

Another example comes from engineer Elmer Barton, who worked at the Cape for most of his career. He recalled an early rocket launch problem. The engineers set up a number of cameras to monitor and film each launch, to learn everything they could from these expensive, brief, and violent events. With just minutes remaining before launch, he saw a truck run over his cables. All the cameras went out, and he heard Mission Control's confused chatter. Barton knew the mission would be aborted any second now if they didn't recover the cameras. A damaged cable sprawled there in front of him but otherwise unseen. "Without saying anything to anybody," he said. "I go down on my knees with my pocketknife and peeled the cable back." He and a colleague started splicing the cable back together, by hand. One by one the cameras started coming back online, to the mystified delight of flight controllers, and the mission moved ahead. Barton said he never had a second thought about it. "Get the thing done," he said.[7]

Responsibility often arrived with little warning. Typically, a young, fresh-faced employee could find herself or himself called to a big meeting, hoping to just listen and learn instead of speaking. Then, when a thorny, unsolved problem arose, a boss invariably stood up, pointed to the shrinking youngster, and introduced them as the project manager for the previously unconsidered issue. In some cases, a manager would say, "The floor is yours," and then excuse himself, saying he was late for another meeting. These engineers, thrown into the deep end, had to tread water before they mastered new subjects, but it usually worked in the end. In most meetings, the group rallied to help outline the way forward.[8]

Henry Pohl and others share similar Huntsville stories. Youngsters would find themselves in the smoke-filled von Braun conference room. Perhaps a boss would be conveying some new finding, and von Braun would suddenly interrupt. "Let *him* tell me about it himself," he'd say, pointing at a youngster. And in this pre-PowerPoint era, with no notes or posters on hand, the blushing engineer would get up and hold forth. By the end of the meeting, he might own a new project—a project that had owned neither name nor acronym just an hour earlier.

DIRTY HANDS

In theory, any manager could empower their troops. But in the NASA of the 1960s, they all had another key quality. Dating from his days with the V-2 rocket, von Braun told his colleagues in Huntsville he wanted his group and laboratory leaders to "keep their knowledge up to date and judgment sharp by keeping their hands dirty at the work bench." And when Robert Gilruth was planning the new space center south of Houston, he told his inner circle they had to build more than office buildings. "I do not want to have our people, our engineers, sit in their offices and only look at paper," he said. "I want them to get their hands dirty, understand the hardware, bring it here and test it." Both Gilruth and von Braun wanted all of their managers to maintain technical chops.[9]

When thinking back on the space program, Henry Pohl starts with the leaders. "Every one of those people grew up in a laboratory doing things themselves," he says. "So when it came time to manage [hundreds of thousands of] people all over the United States and contractors of every persuasion, they could do it from the standpoint of having been there, of knowing what it took to do the job."[10]

A number of engineers tell stories of under-estimating von Braun's own engineering skills. After his magazine and television appearances, some engineers assumed he was mainly a silver-tongued space salesman. But starting from the earliest NASA days, when Henry Pohl asked von Braun about balancing a spinning space station, von Braun could easily dive into tiny technical details, and he expected the same of his managers. "I always thought of von Braun as being a figurehead," Marlowe Cassetti said. But one day late in the Apollo program, Cassetti had to brief key leadership on his study of the lander's weight problem. Von Braun peppered him with penetrating equipment questions. They got down into the ounces of obscure parts in the lander's engine. "I shouldn't have been surprised [that] he knew the technical details."[11]

Engineer Aldo Bordano started at the Houston center straight out of college. "Our bosses were really something we wanted to emulate," he says. "I wanted to be able to ask intelligent questions." He pauses a bit. Thinking back over his entire career, he says it was successful exactly because he spent the entire time trying to embody his NASA elders.[12]

THE TEAMWORK

This might go without saying, but the retired engineers mention goodwill and sacrifice in NASA's early years. "It was just instilled in people," Marlowe Cassetti recalled. "You would see people who would try to build empires," obstructing others, accumulating power, and hoarding resources, as in most any organization. But at NASA "they would be just absolutely steamrolled by this *team*. I saw it time and time again where there was this intense feeling of teamwork." You didn't have to be nice—people were routinely brutal and competitive with one another on technical issues. As Cassetti says, "It wasn't always smooth," but you had to be selfless. Many of the engineers have stories about a person hoarding information or looking out for his own career finding himself pushed effortlessly aside (and often out of the agency). The enormous group spirit, like a rushing flood, removed clutter without expending extra energy.

"You know, I never did worry about who got credit for anything," Henry Pohl says, "as long as the right thing was done." He cites the "common cause" of beating the Russians as the primary welding power at work, convincing the engineers, as individuals, not to worry about the next promotion or their slice of acknowledgment. Some still yearn for those early days, when, as one says, "everybody needed everybody." People worked their insane hours and owned their jobs, but they recognized that other jobs were just as critical, and more often than not, one engineer would drop everything to go help someone else, to lend advice or give brutal feedback on an attempted fix.

Max Faget long partnered with an intuitive and untrained Virginian who'd made his engineering start at Langley. Caldwell Johnson in many ways completed Faget's brilliance. As Faget would dream, Johnson would find a way to get the dream on paper and start making something *possible* out of it. He, like so many others, just saw what needed to happen and was all too happy to fill that gap. He eventually saw the concept of "concinnity" in his life's work. "It's hard to pronounce," he said. "It means a kind of artful assembly of things. That's probably what I did, was artfully assemble things that other people had done into a combination that was pleasant. I didn't invent. . . . It was just a matter of taking the ideas that other people had and putting them together in a way that everybody could look at and say, 'Yes, that's what we want.'"[13]

An artful assembly of people and effort may well have benefited from the striking youth involved. Flexible, open minds abounded—"Why not try it *this* way?" they said in ever-shifting teams. John Mayer, the man who hired my father and scores of other planning engineers for NASA, once said, "I was encouraged by the management to try to hire 'experienced' personnel. But in my mind, I wanted people who didn't know that going to the Moon was an impossible task (as the MIT science advisor to President Eisenhower believed). So, I hired the best of the new college graduates. At one time, I was the oldest one in the division at thirty-six!" Peter Armitage started at a relatively old thirty, and he recalled an army major visiting his office. "He had this puzzled look on his face. He'd been there a day and he'd talked to a lot of our people. . . . And he said, 'Tell me, are there no older people here?' " The military man was used to authority and roles of leadership having a natural link to age.[14]

Waves of people in their twenties could not only sink long hours into an engineering adventure; they were less likely to demand recognition for their ideas, seethe over blunt criticism, or wonder about the next pay raise. Many recall their NASA job offer as the most meager of several salary options. But they wanted to join the space race. As authors Murray and Cox write, parts of NASA had a sort of "male bonding . . . on a grand scale, and a kind of closeness that many of them would never know again." At times, it seemed that the engineers "lacked only decoder rings and a treehouse." If nothing else, most of the engineers will credit their youth for giving them the requisite energy and the sort of fearlessness they needed for the many hurdles they faced.[15]

And fifty years later, the field of neuroscience has made its own giant leaps since Apollo. The brain, we now know, busily sculpts its garden of neural connections well into our twenties. Our thinking is literally more plastic, more flexible, and less fixed in comfortable ruts during our early years. It may well be that when minds needed to change in the face of dead ends and new realities, those in the Apollo program could do so. Brains more stuck in traditional tracks may have spent more time writing up "minority reports" and taking their concerns up the ladder. (And that did happen now and again, to be sure.)

My father came to NASA at age thirty-one, joining with waves of younger engineers. Meanwhile, in the leadership, Faget and company were war veterans in their forties and up. Did my father feel caught in-between, age-wise? Did he fit in? "It never occurred to me," he said, giving the question a reaction usually reserved for odd smells. "I don't think it occurred to anybody." And then he says something they all say, and it has the sound of a deep truth: "We were all just about the work."

BUR-EAU-CRA-WHAT?

Whether springing from the agency's own relative youth or just the immense pressure of their goal, most of the engineers also express nostalgia for the blissful lack of red tape, especially in the early Apollo years.

"The bureaucracy just didn't exist," said Hal Beck, who saw the project from soup to nuts, starting in 1958. "Because we didn't have time and it wasn't tolerated."[16] To be sure, some of this sprang from the roots of Langley's freewheeling research-based culture, but there was even less overhead, by all appearances, in von Braun's Huntsville outfit. They had a "Why wait?" approach to most of their work. For instance, young Henry Pohl once struggled to diagnose a problem within a burning rocket engine.

He needed to somehow *see* it close up, while it was burning. "Henry, this is the army," his German boss Guenther Haukohl said in one of those pre-NASA Huntsville days. "The army has lots of tanks. . . . Go down and get you a tank." So, Pohl rolled a tank up to the rocket stand and watched the engine through its thick glass window.

To be sure, there were management mistakes, and some strategies ended in cul-de-sacs; NASA sometimes backfired and suffered near implosions of organization. During the more intense waves of hiring, some new people said they felt lost and underused. And the earliest days of the Houston center may actually have suffered from the Langley-based, free-flowing structure. Pohl, on transferring to Houston, encountered some frustration. He thought a lot of their meetings and paperwork seemed fairly useless. "I['ll] tell you what," he says now. "Max [Faget] never did think too much of an organization. He was a kind of dreamer. He had the philosophy that if it was the will of the people, any organization would work." And how about Bob Gilruth, the father of manned spaceflight? "Had the same problem," Pohl says.[17] But both Gilruth and Faget eventually recognized the importance of management and hired excellent deputies to apply structure. The efforts in Houston also benefited from new leadership sweeping into power in 1963, when NASA floundered in bad press and constant doubt. Moving from industry, George Mueller, NASA's new administrator for manned spaceflight, erected a scaffolding of logical professionalism—a new "systems engineering" approach.

A number of engineers highlight what they accomplished after the tragic launch pad fire in early 1967. "It couldn't be done today," engineer Thomas Moser said. "You could not do that kind of redesign without having so many checks and balances in the system. It would take years to do it. I think we did it, what, in eight months? . . . from complete redesign to flying again. That was, I think, indicative of the can-do, will-do, and allowed-to-do environment."[18]

Marlowe Cassetti, enjoying front row seats to the space age from the beginning, acknowledged the bureaucracy growing even during the late 1960s. He said it got harder and harder to simply get things done at NASA after the early days of Mercury (phase I) and Gemini (phase II). "On Apollo, much more was being done in committees," he says. "It was more headaches." Maybe we can add bureaucracy to the inevitability club with death and taxes. Bureaucracy caught up with NASA in a hurry, and most voices point to 1970 as the year it became obvious to all. "The manned space program started with one hundred and fifty people," said astronaut Gordon Cooper, noting how it had grown to tens of thousands of agency employees in just ten years. "Typical bureaucracy set in, in a big way." He attributed all the success to this early lean structure. "The way we caught the Russians was by not really having a bureaucracy to start with. Then, you could . . . make a total change in fifteen minutes, because it was all first-name acquaintance. You scribbled out a little piece of paper, handed it around, and the change was made." Just a few years later, "that same change would take you a month!"[19]

Again, we see the arc: NASA launched its incredible journey from 1958, leaving everyday, grounded concerns behind with billowing exhaust. When the agency came back to Earth after Apollo, it saw not just a completely changed America, but when it looked in the mirror it saw an aged and bloated version of itself: successful and rightly proud, but hardly ready to take on a new marathon.

Some of those from Apollo leadership speak angrily of the lost momentum. "If I'd had a clue that it would be forty or fifty more years before we sent men back there," Chris Kraft wrote, "I would have fought like hell to get Apollos 18 through 20 back into the schedule." Kraft disciple and flight director Gerry Griffin feels the same way. He now often gives talks to schools about what it will take next time we get the chance to explore. "I tell them, 'you've got to be ready, because these factors *will* line up again.'"[20]

Despite their pride in Apollo, some voices from the trenches might not re-enlist. "If I knew then what I know now," Caldwell Johnson, Max Faget's right-hand man, said in 1999, "I think I'd have gone to Australia or somewhere and got the hell out of there."

And the biggest surprise comes from Henry Pohl: "I would not in any way, shape, size, or form want to go through the Apollo Program [again]," he said. "But that's all I did for seven years. That's all I thought about, day and night. That's all I dreamed about, day and night. We had so many problems, and you never had enough time." He said a lot of it was fun, but it took too much of a toll. "I guess I remember in '65 making fifty-one trips out of Houston, and the reason I remember that is there's fifty-two weeks in a year." With his Houston desk disappearing in piles of worries, Pohl had tracked insolvent problems, arguments, and mysteries from Huntsville and the Cape to dozens of contractors. He could not afford to ignore a single loose thread.

Reflecting on Apollo, I keep thinking of my father's lifelong fascination with the aboriginal history of Texas. With a cousin of mine—a moonlighting archeologist—he's spent many hours hunting arrowheads and even visiting a few of the precious raised aboriginal mounds. He pursued flint knapping for several years and has recently returned to it. By bringing rock pieces clacking together over meditative hours, he eventually holds a convincing arrowhead of his own. The warm flint smells like a campfire from centuries past.

He's no anthropologist, but my father likes to talk of mound building as a natural human drive. Whether it's a pyramid in northern Africa or a temple in Central America, or a larger-than-needed ceremonial mound in pre-Columbian Texas, given full stomachs and a little spare time, it seems that many human cultures drive themselves to collective efforts as large as they are whimsical. These are not prag-matic outbursts from human groups. The projects often appear useless in terms of survival.

And my father has always seen Apollo in the same light: a cultural peak of in-spired collaboration, driven by a mix of curiosity, vision, and irrational fear, instead of by hunger or exposure to the elements. As astronaut and physicist Joseph Allen put it, Apollo was "the aggregate of virtually every bit of human skill and knowledge," from computing and transmitting to welding and sewing. That makes it a rarified and precious act. But as many others have noted, it's impossible to ascribe meaning to Apollo. Every time someone tries, the resulting verbiage tumbles out disheveled. The words sound vapid, incomplete, vague, or cliché.[21] In the end, perhaps it was this simple: A child-like species decided it would try to leap and touch a branch of heaven that once seemed impossibly high.

Fifty years later, surviving engineers tend to avoid surrounding it with meaning. Apollo did not necessarily start a sustained colonization of the cosmos. It did not bring world peace or permanently reform the myopic ways of humankind. Maybe Apollo's memory will inspire a future culture to go further into the inhospitable uni-verse, or maybe the program will be forgotten or even denied. To many engineers, it's just something they leave behind, their built mound. The data tapes, photographs, and blueprints will all eventually fade and crumble here on Earth. But on the Moon's stark landscape, the retired rovers and leggy landing stages will abide for millions of years to come.

NOTES

Preface

1. McDougall, . . . *the Heavens and the Earth*, pp. 7, 140, 145; Murray and Cox, *Apollo*, p. 24.

Chapter 1

1. Lasby, *Project Paperclip*, p. 26; Neufeld, *Von Braun*, pp. 159–160, 190, 61–62.
2. Von Braun background thanks primarily to: Neufeld, *Von Braun*, Chapters 2 and 7; Ward, *Dr. Space*, Chapter 6; DeGroot, *Dark Side of the Moon*, pp. 20–21; and Piszkiewicz, *Wernher von Braun*, Chapter 1.
3. Piszkiewicz, *Wernher von Braun*, pp. 5–6, 8; Ward, *Dr. Space*, p. 65.
4. Ward, *Dr. Space*, pp. 55–59; NASA Oral Histories interview with Ernst Stuhlinger.
5. Gainor, *To a Distant Day*, pp. 43–44, 58, 19–22; Neufeld, *Von Braun*, pp. 24–25.
6. Piszkiewicz, *Wernher von Braun*, pp. 1–3; DeGroot, *Dark Side of the Moon*, p. 17.
7. Kennedy, Henderson, and Siley, *Prelude to Leadership*, pp. 6–8.
8. Interview with Guy Faget, 2017, and official naval records of the USS Guavina.
9. Interview with Carol Faget, 2016.
10. Murray and Cox, *Apollo*, pp. 40–41.
11. Interviews with Henry Pohl, 2016 and 2017, and unpublished written recollections of Henry Pohl.
12. Interviews with Marlowe Cassetti, 2016 and 2017.
13. Gainor, *To a Distant Day*, pp. 41, 58–60, 63–64.

Chapter 2

1. Neufeld, *Von Braun*, p. 27.
2. Werner von Braun amalgam here via Neufeld, *Von Braun*, pp. 256–262; Kraft, *Flight*, pp. 62–63; Piszkiewicz, *Werner von Braun*, pp. 74–78, 100; and Ward, *Dr. Space*, Chapters 9, 10, and 13.
3. McDougall, . . . *the Heavens and the Earth*, p. 224.
4. Interviews with Henry Pohl, 2016 and 2017; unpublished written recollections of Henry Pohl.
5. Oberg, *Red Star in Orbit*, p. 25.
6. Oberg, *Red Star in Orbit*, p. 30; Reeves, *The Superpower Space Race*, p. 18.
7. Dickson, *Sputnik*.
8. Oberg, *Red Star in Orbit*, p. 32.
9. Launius, Logsdon, and Smith, *Reconsidering Sputnik*, pp. 270–271.
10. McDougall, . . . *the Heavens and the Earth*, pp. 140, 145.

11. As reprinted in Launius, *NASA*, pp. 144–146.
12. Interviews with Robert Austin, 2017.
13. NASA Oral Histories interview with Ernst Stuhlinger.
14. McDougall, . . . *the Heavens and the Earth*, p. 150; Gainor, *To a Distant Day*, pp. 119–120.
15. Ward, *Dr. Space*, pp. 112–113.
16. NASA Oral Histories interview with Marlowe Cassetti.
17. NASA Oral Histories interview with Joseph "Guy" Thibodaux, who was Faget's friend in the car; 2016 interview with Henry Pohl.
18. NASA Oral Histories interview with Max Faget.
19. Murray and Cox, *Apollo*, pp. 39–40.
20. NASA Oral Histories interview with Max Faget.
21. Kraft, *Flight*, p. 63.
22. NASA Oral Histories interview with Guy Thibodaux.
23. McDougall, . . . *the Heavens and the Earth*, p. 154.

Chapter 3

1. Dunar and Waring, *Power to Explore*, p. 52.
2. Interview with Marlowe Cassetti, 2017.
3. Hal Beck, unpublished historical writing.
4. Interview with Marlowe Cassetti, 2018; Ward, *Dr. Space*, pp. 120–121.
5. Murray and Cox, *Apollo*, pp. 51–52.
6. Bilstein, *Stages to Saturn*, Chapter 3.
7. Dunar and Waring, *Power to Explore*, p. 84.
8. Interviews with Thomas Parnell and Nesbitt Cumings, 2017.
9. NASA Oral Histories interview with Henry Pohl; interview with Karl Pohl, 2017; interview with Henry Pohl, 2017; Marshall Space Flight Center film of Saturn model, 1959, courtesy Henry Pohl.
10. Piszkiewicz, *Wernher von Braun*, pp. 138–139; Oberg, *Red Star in Orbit*, pp. 39–51.
11. McDougall, . . . *the Heavens and the Earth*, p. 222.
12. Kraft, *Flight*, p. 351.
13. Von Ehrenfried, *The Birth of NASA*, p. 316.
14. Shetterly, *Hidden Figures*, pp. xvi–xvii, 138–139, 171; interview with Catherine Osgood, 2016; NASA Oral Histories interview with Catherine Osgood.
15. Shetterly, *Hidden Figures*, pp. 135, 168, 181–185.
16. Interview with Catherine Osgood, 2016; NASA Oral Histories interview with Catherine Osgood.
17. NASA Oral Histories interview with Caldwell Johnson and Robert Chilton.
18. Launius and Jenkins, *Coming Home*, pp. 16–18, 26–28.
19. NASA Oral Histories interview with Aleck Bond.
20. Launius and Jenkins, *Coming Home*, pp. 71–72.
21. Newport, *Lost Spacecraft*, p. 49; NASA Oral Histories interviews with Peter Armitage and Robert Chilton.
22. Ward, *Dr. Space*, p. 84; NASA Oral Histories interview with Henry Pohl.
23. Von Ehrenfried, *The Birth of NASA*, p. 56; NASA Oral Histories interview with Ronald Rose.
24. NASA Oral Histories interview with Owen Maynard.

25. Smithsonian interviews with Robert Gilruth.
26. Kraft, *Flight*, pp. 118–119.

Chapter 4

1. NASA Oral Histories interview with Max Faget.
2. Interview with Ken Young (re: rendezvous), 2016.
3. McDougall, . . . *the Heavens and the Earth*, pp. 229–230.
4. Murray and Cox, *Apollo*, p. 69; de Monchaux, *Spacesuit*, p. 143; Brooks, Crimwood, and Swenson, *Chariots for Apollo*, p. 15.
5. Kraft, *Flight*, pp. 124–126; NASA Oral Histories interview with Peter Armitage.
6. Respectively, Oberg, *Red Star in Orbit*, pp. 52–54, and Kraft, *Flight*, p. 131.
7. Following de Monchaux's *Spacesuit*, pp. 132–133, 144.
8. NASA Oral Histories interview with Arnold Aldridge.
9. de Monchaux, *Spacesuit*, p. 144; Johnson statement via McDougall, . . . *the Heavens and the Earth*, p. 8; NASA Oral Histories interview with Robert Gilruth; Ward, *Dr. Space*, Chapter 14. For further drilling see Logsdon, *The Decision to Go to the Moon*.
10. Scott and Jurek, *Marketing the Moon*, p. 112.
11. Steigerwald, *The Sixties and the End of Modern America*, pp. 101, 13.
12. Ward, *Dr. Space*, pp. 127–128.
13. NASA Oral Histories interviews with Henry Pohl, Caldwell Johnson, and Marlowe Cassetti.
14. Interview with Marlowe Cassetti, 2016.
15. Bilstein, *Stages to Saturn*, Chapter 9.
16. NASA Oral Histories interview with Henry Pohl.
17. Ward, *Dr. Space*, pp. 101–108; interview with Don Woodruff, 2017; interviews with Thomas Parnell, 2016 and 2017.
18. NASA Oral Histories interview with Marlowe Cassetti; Dethloff, *Suddenly Tomorrow Came*, pp. 38–40.
19. De Groot, *Dark Side of the Moon*, p. 170.
20. Interview with Marlowe Cassetti, 2016; NASA Oral Histories interview with Hal Beck.
21. NASA Oral Histories interview with Aleck Bond.
22. NASA Oral Histories interview with Caldwell Johnson.
23. Kraft, *Flight*, pp. 351, 83.
24. Recounted by Henry Pohl in a 2017 interview.
25. Benson and Faherty, *Moonport*, p. 4; "V-2, off Course, Falls near Juarez," *El Paso Times*, May 30, 1947.
26. Wolfe, *The Right Stuff*, pp. 162–163.
27. Interview with Don Woodruff, 2017.
28. Murray and Cox, *Apollo*, pp. 61–62.
29. Murray and Cox, *Apollo*, pp. 61–62; Benson and Faherty, *Moonport*, p. 37; Seamans, *Project Apollo*, p. 6.
30. Seamans, *Project Apollo*, pp. 34, 39.
31. Tsiao, *"Read You Loud and Clear,"* pp. 30–32.
32. NASA Oral Histories interview with Arnold Aldridge; Tsiao, *Read You Loud and Clear*, pp. 35–36.
33. Interview with Marlowe Cassetti, 2016.

34. I am indebted to Hal Beck for sharing his documentation of the early manned space program.

35. Interview with Marlowe Cassetti, 2016.

36. Credit to Tribbe's *No Requiem for the Space Age*, where this idea is a central, convincing theme.

37. Kraft, *Flight*, pp. 153–154.

Chapter 5

1. Rothery, *Moons*, pp. 17–59; for the stabilization argument, see, for instance, Ward and Brownlee, *Rare Earth*, pp. 221–223; for the tidal pool limb development argument, the idea traces to paleontologist Alfred Romer in pre-Apollo days, but for more recent developments see, for instance, the work of astrophysicist Steven Balbus and his collaborators.

2. Compton, *Where No Man Has Gone Before*, p. 27; Beattie, *Taking Science to the Moon*, p. 18.

3. NASA Oral Histories interview with Caldwell Johnson; Beattie, *Taking Science to the Moon*, p. 15; Pfeiffer, "Moon Mysteries Ranger Didn't Solve."

4. NASA Oral Histories interview with Caldwell Johnson.

5. Interviews with Hal Beck and Frank Hughes, 2018; NASA Oral Histories interview with Frank Hughes.

6. Reeves, *The Superpower Space Race*, pp. 46–48; NASA Oral Histories interview with Frank Hughes.

Chapter 6

1. Tribbe, *No Requiem for the Space Age*, pp. 162, 298–299.

2. Kraft, *Flight*, p. 156.

3. Interview with Marlowe Cassetti, 2016; NASA Oral Histories interview with Kenneth Kleinknecht.

4. NASA Oral Histories interview with Marlowe Cassetti.

5. Von Ehrenfried, *The Birth of NASA*, p. 176; Wolfe, *The Right Stuff*, p. 350.

6. Interview with Carol Faget, 2016.

7. NASA Oral Histories interview with Aleck Bond; von Ehrenfried, *The Birth of NASA*, p. 93.

8. As in Brooks, Crimwood, and Swenson, *Chariots for Apollo*, pp. 61–64.

9. Interview with Thomas Moser, 2016.

10. John Mayer's analogy as relayed in Murray and Cox, *Apollo*, pp. 327–328.

11. NASA Oral Histories interview with Hal Beck.

12. Interview with Marlowe Cassetti, 2016; NASA Oral Histories interview with Hal Beck; interview with Hal Beck.

13. Interview with Ken Young, 2016; Mindell, *Digital Apollo*, pp. 125, 127.

14. Interview with Marlowe Cassetti, 2016.

15. Primarily following Bilstein, *Stages to Saturn*, Chapter 4; NASA Oral Histories interview with Joseph "Guy" Thibodaux; Kelly, *Moon Lander*, p. 133.

16. Dunar and Waring, *Power to Explore*, pp. 89–90; Bilstein, *Stages to Saturn*, Chapter 7.

17. Bilstein, *Stages to Saturn*, Chapter 5.

18. Murray and Cox, *Apollo*, p. 98; Benson and Faherty, *Moonport*, p. 118.
19. Dunar and Waring, *Power to Explore*, p. 62.
20. Ward, *Dr. Space*, pp. 91, 170.
21. Interview with Henry Pohl, 2017.
22. Unpublished written recollections of Henry Pohl.
23. Murray and Cox, *Apollo*, pp. 148–151; Bilstein, *Stages to Saturn*, Chapter 4.

Chapter 7

1. Jastrow and Newell, "Why Land on the Moon?" *Atlantic Monthly*, August, 1963.
2. Murray and Cox, *Apollo*, pp. 152–153.
3. Fries, *NASA Engineers and the Age of Apollo*, p. 126; NASA SP-4214, pp. 30–31; Kraft, *Flight*, pp. 212–213; 349–350.
4. Brooks, Crimwood, and Swenson, *Chariots for Apollo*, p. 365; NASA Oral Histories interview with Harold Beck.
5. Oberg, *Red Star in Orbit*, pp. 67–71.
6. de Monchaux, *Spacesuit*, pp. 154–155.
7. Wolfe, *The Right Stuff*, p. 435; McDougall, . . . *the Heavens and the Earth*, p. 394; Piszkiewicz, *Wernher von Braun*, p. 153; S. Khrushchev in Launius, Logsdon, and Smith, *Reconsidering Sputnik*, p. 283.
8. Poignant technological detail courtesy of McDougall, . . . *the Heavens and the Earth*, p. 396.
9. Reeves, *The Superpower Space Race*, p. 67.
10. Oberg, *Red Star in Orbit*, p. 76.
11. Neufeld, *Von Braun*, p. 405.
12. Ward, *Dr. Space*, pp. 77–78.
13. Steigerwald, *The Sixties and the End of Modern America*, p. 23.
14. Kelly, *Moon Lander*, pp. 134–135; Bilstein, *Stages to Saturn*, Chapter 4. The contractor in question was Grumman, and the engineers were traveling to Bell Labs.
15. Bilstein, *Stages to Saturn*, Chapter 4; Dunar and Waring, *Power to Explore*, p. 97.
16. Interview with Len Worlund, 2017; Murray and Cox, *Apollo*, p. 313.
17. Interview with Len Worlund, 2017.
18. Interview with Bill Sneed, 2017.
19. NASA Oral Histories interview with Henry Pohl; interview with Henry Pohl, 2017.
20. Ward, *Dr. Space*, pp. 149–150.
21. Brooks et al., *Chariots for Apollo*, p. 165.
22. Kelly, *Moon Lander*, p. 93.
23. NASA Oral Histories interview with Max Faget.
24. Kelly, *Moon Lander*, p. 63. Quote attributed to Tom Kelly himself.
25. Various Faget attributions in this section are from the NASA oral histories interviews.
26. Mindell, *Digital Apollo*, p. 183; Brooks et al., *Chariots for Apollo*, pp. 244–247.
27. Brooks et al., *Chariots for Apollo*, pp. 146–147; Kelly, *Moon Lander*, p. 66.
28. Kelly, *Moon Lander*, p. 132.
29. Murray and Cox, *Apollo*, p. 336.
30. Kelly, *Moon Lander*, p. 82.
31. De Monchaux, *Spacesuit*, p.189.
32. De Monchaux, *Spacesuit*, p. 239.

33. De Monchaux, *Spacesuit*, pp. 189, 211, 209.
34. Interview with Aldo Bordano, 2016.
35. Interview with Lee Norbraten, 2016.
36. Interview with Aldo Bordano, 2016.
37. Interviews with Lee Norbraten and Mack Henderson, 2016.
38. Murray and Cox, *Apollo*, p. 173.
39. NASA document SP-4214; Brooks et al., *Chariots for Apollo*, p. 204.
40. Interview with Henry Pohl, 2016.
41. Interviews with Ann Faget and Carol Faget, 2016; interview with Guy Faget, 2017.
42. Interview with Marlowe Cassetti, 2018; Kurlansky, *1968*, p. 89; Dunar and Waring, *Power to Explore*, pp. 123–124.
43. The Surfaris, "Wipe Out" (on the album of the same name), Dot Records, 1963.
44. Dunar and Waring, *Power to Explore*, p. 123; Ward, *Dr. Space*, p. 173; Dunar and Waring, *Power to Explore*, pp. 121–122.
45. Dunar and Waring, *Power to Explore*, pp. 116, 118–119.
46. Kraft, *Flight*, p. 205.

Chapter 8

1. NASA Oral Histories interview with Henry Pohl; interviews with Henry Pohl, 2016 and 2017.
2. Van Nimmen and Bruno, *NASA Historical Data Book, Volume 1*, pp. 148–155; Dunar and Waring, *Power to Explore*, p. 46.
3. Dunar and Waring, *Power to Explore*, p. 79.
4. Interview with Len Worlund, 2017.
5. Interviews with Len Worlund and Nesbitt Cumings, 2017.
6. Interview with Robert Austin, 2016.
7. Dunar and Waring, *Power to Explore*, pp. 89–90; Bilstein, *Stages to Saturn*, Chapter 7.
8. Newport, *Lost Spacecraft*, p. 53; Brooks, Crimwood, and Swenson, *Chariots for Apollo*, pp. 239–240.
9. Oberg, *Red Star in Orbit*, pp. 79–81; NASA Oral Histories interview with Larry Bell; de Monchaux, *Spacesuit*, p. 112.
10. De Monchaux, *Spacesuit*, pp. 192–194.
11. NASA Oral Histories interview with Alec Bond.
12. NASA Oral Histories interview with Caldwell Johnson.
13. Interview with Marlowe Cassetti, 2018. He recalls Canadian imports Richard Carley and Richard Chamberlain being two of the chief proponents of a Gemini Moon mission.
14. NASA Oral Histories interview with Ernst Stuhlinger; Brooks et al., *Chariots for Apollo*, pp. 180–181.
15. Kraft, *Flight*, 228–233.
16. French and Burgess, *In the Shadow of the Moon*, p. 34.
17. Interview with Cynthia Wells, 2016.
18. Interview with Wesley Ratcliff, 2017.
19. Interview with Thomas Moser, 2016.
20. NASA Oral Histories interview with Cathy Osgood.

21. NASA Oral Histories interview with Ken Young; Tsiao, *Read You Loud and Clear*, pp. 109, 117–127, 141; interview with Ken Young, 2016.
22. NASA Oral Histories interview with Elmer Barton.
23. Kraft, *Flight*, p. 240; Astronaut Tom Stafford, as quoted in French and Burgess, *In the Shadow of the Moon*, p. 75.
24. Kraft, *Flight*, pp. 242–243.
25. NASA Oral Histories interview with Ken Young. This story could also have come from Gemini V, which was also reportedly quite fragrant.
26. Murray and Cox, *Apollo*, p. 181; Benson and Faherty, *Moonport*, p. 431; Murray and Cox, *Apollo*, pp. 94, 247–248; Benson and Faherty, *Moonport*, pp. 282–283.

Chapter 9

1. NASA Oral Histories interview with Jack Garman.
2. Mindell, *Digital Apollo*, pp. 139–140.
3. NASA Oral Histories interview with Jack Garman.
4. Mindell, *Digital Apollo*, pp. 143, 151–152.
5. Mindell, *Digital Apollo*, pp. 172, 169.
6. Mindell, *Digital Apollo*, p. 154; NASA Oral Histories interviews with Jack Garman and Robert Chilton; interview with Frank Hughes, 2017.
7. Oberg, *Red Star in Orbit*, p. 85; Harford, *Korolev*, pp. 277–282.
8. Reeves, *The Superpower Space Race*, pp. 101–103, 105.
9. Brooks, Crimwood, and Swenson, *Chariots for Apollo*, p. 206; Reeves, *The Superpower Space Race*, pp. 111, 113–114.
10. Kelly, *Moon Lander*, p. 104.
11. Interview with Thomas Moser, 2016.
12. NASA Oral Histories interview with Alec Bond.
13. Interview with Frank Hughes, 2016. The incident described may well be one from August of 1966, as discussed in Benson and Faherty, *Moonport*, p. 344.
14. Benson and Faherty, *Moonport*, pp. 485–486.
15. NASA Oral Histories interviews with Frank Hughes.
16. NASA Oral Histories interviews with Frank Hughes; de Monchaux, *Spacesuit*, p. 171.
17. French and Burgess, *In the Shadow of the Moon*, pp. 83–85; Kraft, *Flight*, pp. 254–256; interview with Henry Pohl, 2017.
18. Henry Pohl, unpublished written recollections.
19. NASA Oral Histories interview with Henry Pohl; interview with Henry Pohl, 2017; interview with Aldo Bordano, 2016.
20. Kraft, *Flight*, pp. 258–260; French and Burgess, *In the Shadow of the Moon*, pp. 98–103.
21. Bilstein, *Stages to Saturn*, Chapters 3 and 7. The Mississippi test facility later became the Stennis Space Center.
22. Interview with William Sneed, 2017.
23. Dunar and Waring, *Power to Explore*, pp. 89–90.
24. Murray and Cox, *Apollo*, p. 237.
25. Ward, *Dr. Space*, pp. 158–159; Piszkiewicz, *Wernher von Braun*, p. 181.
26. Farber, *The Sixties*, p. 30; Compton, *Where No Man Has Gone Before*, pp. 48–49.
27. Kurlansky, *1968*, pp. 96, 114–115.
28. Steigerwald, *The Sixties*, overall, but pp. 243–244 in particular here.

29. NASA Oral Histories interview with Henry Pohl.
30. Murray and Cox, *Apollo*, pp. 185–186. The vice president was Hilliard W. Paige of General Electric.
31. Reeves, *Superpower Space Race*, pp. 120–121.

Chapter 10

1. NASA Oral Histories interviews with Caldwell Johnson and Max Faget.
2. Oberg, *Red Star in Orbit*, pp. 112–113, 121–123; Harford, *Korolev*, 288–293.
3. Murray and Cox, *Apollo*, pp. 186–187.
4. Brooks, Crimwood, and Swenson, *Chariots for Apollo*, p. 214; Murray and Cox, *Apollo*, pp. 186–187; Kraft, *Flight*, pp. 269–270.
5. Murray and Cox, *Apollo*, p. 197.
6. Ibid., p. 220.
7. NASA Oral Histories interview with Larry Bell; "Second Airman Dies from Spaceship [*sic*] Burns," *Madera Tribune*, February 1, 1967. The airmen were William Bartley Jr. and Richard Harmon.
8. Murray and Cox, *Apollo*, p. 203.
9. Kraft, *Flight*, p. 275; Kelly, *Moon Lander*, p. 161.
10. Kraft, *Flight*, p. 276.
11. Ward, *Dr. Space*, p. 140; Reeves, *The Superpower Space Race*, pp. 123–124.
12. NASA Oral Histories interview with Henry Pohl.
13. Ibid.
14. Kelly, *Moon Lander*, pp. 137, 122.
15. NASA Oral Histories interviews with Marlowe Cassetti and Thomas Moser.
16. Bilstein, *Stages to Saturn*, Chapter 6.
17. Bilstein, *Stages to Saturn*, Chapter 12; Murray and Cox, *Apollo*, pp. 239–240.
18. Rabinowitch and Lewis, *Man on the Moon*, pp. 57–58.
19. Benson and Faherty, *Moonport*, pp. 281–282.
20. The quote is from Grady Corn, as in Murray and Cox, *Apollo*, p. 241.
21. Interview with Gerry Griffin, 2016.
22. Bilstein, *Stages to Saturn*, Chapter 8.
23. Ibid.
24. Tribbe, *No Requiem for the Space Age*, pp. 198, 170.
25. Benson and Faherty, *Moonport*, p. 428.
26. Murray and Cox, *Apollo*, p. 248; Kraft, *Flight*, p. 279.
27. Benson and Faherty, *Moonport*, p. 429; Kraft, *Flight*, p. 278; Dunar and Waring, *Power to Explore*, p. 95; Murray and Cox, *Apollo*, pp. 249–250.
28. Bilstein, *Stages to Saturn*, Chapter 4; Mindell, *Digital Apollo*, p. 175.
29. Dunar and Waring, *Power to Explore*, p. 142.

Chapter 11

1. Murray and Cox, *Apollo*, p. 308; interview with Karl Pohl, 2017.
2. Kurlansky, *1968*, pp. 114–115; Cleaver, *Soul on Ice*, cited in Kurlansky, *1968*, p. 345.
3. De Monchaux, *Spacesuit*, p. 11; Scott and Jurek, *Marketing the Moon*, p. 25.
4. Tribbe, *No Requiem for the Space Age*, pp. 3–5.

5. Tribbe, *No Requiem for the Space Age*, pp. 209–210.

6. Farber, *The Sixties*, p. 32; Harford, *Korolev*, p. 291.

7. Murray and Cox, *Apollo*, pp. 309–310; Brooks, Crimwood, and Swenson, *Chariots for Apollo*, pp. 247–249.

8. Mindell, *Digital Apollo*, p. 176; NASA Oral Histories interview with Peter Armitage.

9. As quoted in the introduction to Watkins, *Apollo Moon Missions*.

10. Reeves, *The Superpower Space Race*, p. 129; Mindell, *Digital Apollo*, p. 213.

11. Following Murray and Cox, *Apollo*, pp. 313–314; for the pogo fix, following Dunar and Waring, *Power to Explore*, p. 97.

12. Ward, *Dr. Space*, pp. 158–159. Quote originally from October 1968 *US News and World Report*.

13. Brooks et al., *Chariots for Apollo*, p. 256.

14. NASA Oral Histories interview with Marlowe Cassetti; interview with Marlowe Cassetti, 2017.

15. Brooks et al., *Chariots for Apollo*, p. 255.

16. NASA Oral Histories interview with Henry Pohl; interview with Henry Pohl, 2016.

17. Following Kraft's account in *Flight*, pp. 284–285.

18. Interview with Mac Henderson, 2016.

19. Brooks et al., *Chariots for Apollo*, p. 259.

20. Interview with Lee Norbraten, 2016.

21. NASA Oral Histories interview with Frank Hughes.

22. Harford, *Korolev*, pp. 292–293.

23. French and Burgess, *In the Shadow of the Moon*, pp. 213–214; burping story from NASA Oral Histories interview with Ken Young; interview with Ken Young, 2016.

24. Kraft, *Flight*, pp. 294–296.

25. Harford, *Korolev*, p. 292; DeGroot, *Dark Side of the Moon*, p. 227.

26. French and Burgess, *In the Shadow of the Moon*, pp. 303–304.

27. NASA Oral Histories interview with Gerry Griffin; interview with Gerry Griffin, 2016.

28. Interview with Frank Hughes, 2016.

29. French and Burgess, *In the Shadow of the Moon*, pp. 355–357.

30. French and Burgess, *In the Shadow of the Moon*, p. 308.

31. Interview with Frank Hughes, 2017.

32. NASA Oral Histories interview with Frank Hughes; interview with Frank Hughes, 2017.

33. NASA Oral Histories interview with Jack Garman.

34. Interview with Edward Kowalchuk, 2016.

35. NASA Oral Histories interview with Frank Hughes.

36. Tsiao, *"Read You Loud and Clear,"* pp. 147–148, 172.

37. Brooks et al., *Chariots for Apollo*, pp. 278–279.

38. NASA Oral Histories interview with Gerry Griffin; Mindell, *Digital Apollo*, pp. 177–178.

39. Murray and Cox, *Apollo*, p. 328.

40. Interview with Hal Beck, 2016.

41. Watkins, *Apollo Moon Missions*, p. 63. Quotes from Simon Bourgin, then a science advisor to the U.S. Information Agency. Joseph Laitin determined the passage with his wife's input.

42. Murray and Cox, *Apollo*, p. 331.

43. French and Burgess, *In the Shadow of the Moon*, p. 315; Mindell, *Digital Apollo*, pp. 178–179, 160.

44. French and Burgess, *In the Shadow of the Moon*, p. 316; Brooks et al., *Chariots for Apollo*, pp. 283–284; NASA Oral Histories interview with Frank Hughes.

45. Ward, *Dr. Space*, p. 139.

46. Kluger, *Apollo 8*, p. 287.

47. Tribbe, *No Requiem for the Space Age*, pp. 130–132.

Chapter 12

1. Interview with Guy Faget, 2017; interview with Henry Pohl, 2017.

2. Neufeld, *Von Braun*, pp. 417, 420, 434–436.

3. Neufeld, *von Braun*, pp. 414–415; interview with Guy Faget, 2017.

4. French and Burgess, *In the Shadow of the Moon*, p. 339; Kelly, *Moon Lander*, p. 124.

5. Brooks, Crimwood, and Swenson, *Chariots for Apollo*, pp. 286–287.

6. French and Burgess, *In the Shadow of the Moon*, pp. 271–274.

7. Oberg, *Red Star in Orbit*, pp. 98–99; French and Burgess, *In the Shadow of the Moon*, pp. 276–277.

8. De Monchaux, *Spacesuit*, pp. 303, 306–307.

9. Tribbe, *No Requiem for the Space Age*, pp. 158–159.

10. Kraft, *Flight*, pp. 303–305; French and Burgess, *In the Shadow of the Moon*, p. 342.

11. French and Burgess, *In the Shadow of the Moon*, pp. 372, 377–378.

12. Combining Apollo 10 accounts: Brooks et al., *Chariots for Apollo*, pp. 303–312; Kraft, *Flight*, pp. 309–310; Kelly, *Moon Lander*, p. 206; and French and Burgess, *In the Shadow of the Moon*, pp. 372–378.

13. Interview with Karl Pohl, 2017.

14. Ward, *Dr. Space*, Chapter 16.

15. Scott and Jurek, *Marketing the Moon*, pp. 80, 83; de Monchaux, *Spacesuit*, pp. 253, 255, 257.

16. NASA Oral Histories interview with Rodney Rose.

17. Murray and Cox, *Apollo*, p. 346; NASA Oral Histories interview with Jack Garman.

18. NASA Oral Histories interview with Thomas Moser.

19. Murray and Cox, *Apollo*, p. 355; Tribbe, *No Requiem for the Space Age*, p. 166.

20. Mindell, *Digital Apollo*, pp. 190–191.

21. NASA Oral Histories interviews with Jack Garman and Ken Young; Watkins, *Apollo Moon Missions*, p. 7; interview with Frank Hughes, 2018.

22. Mindell, *Digital Apollo*, pp. 227–228, 222.

23. Interview with Frank Hughes, 2018; NASA Oral Histories interview with Marlowe Cassetti.

24. Jack Garman, quoted in Murray and Cox, *Apollo*, p. 352; NASA Oral Histories with Frank Hughes; interview with Frank Hughes, 2017.

25. Kelly, *Moon Lander*, p. 222; Murray and Cox, *Apollo*, pp. 382–383.

26. Kelly, *Moon Lander*, pp. 212–213.

27. NASA Oral Histories interviews with Ernst Stuhlinger and Thomas Moser.

28. De Monchaux, *Spacesuit*, pp. 252–253; Kraft, *Flight*, pp. 307–308; Brooks et al., *Chariots for Apollo*, pp. 328–329.

29. De Monchaux, *Spacesuit*, p. 251; Tribbe, *No Requiem for the Space Age*, pp. 27–28.

30. Interview with Chip Lord, 2017.

31. Kelly, *Moon Lander*, p. 216.

32. Delp et al., "Apollo Lunar Astronauts Show Higher Cardiovascular Disease Mortality."

33. Interview with Henry Pohl, 2017; Neufeld, *Von Braun*, p. 434; NASA Oral Histories interview with Peter Armitage; Beattie, *Taking Science to the Moon*, pp. 262–263.
34. Rabinowitch and Lewis, *Man on the Moon*, p. 166; Scott and Jurek, *Marketing the Moon*, p. 38.
35. Interview with Mac Henderson, 2016.
36. Interview with Don Woodruff, 2017.
37. Murray and Cox, *Apollo*, pp. 373–375; interview with Gerry Griffin, 2016. The young engineer who recognized the noise pattern was John Aaron.
38. Mindell, *Digital Apollo*, pp. 237–238; Kelly, *Moon Lander*, p. 222. The engineer perfecting the landing routine was Emil Schiesser.
39. Scott and Jurek, *Marketing the Moon*, p. 69.
40. Interviews with Wesley Ratcliff and Thomas Parnell, 2017.
41. Scott and Jurek, *Marketing the Moon*, p. 70; Steigerwald, *The Sixties*, p. 290.
42. Selvin, Joel. *Altamont: The Rolling Stones, the Hells Angels, and the Inside Story of Rock's Darkest Day*. Harper Collins, 2016.
43. Tribbe, *No Requiem for the Space Age*, pp. 134, 175, 207.

Chapter 13

1. Combining sources: Kraft, *Flight*, pp. 337–339; Murray and Cox, *Apollo*, pp. 389, 391; Benson and Faherty, *Moonport*, p. 487; NASA Oral Histories interviews with Joseph "Guy" Thibodaux, Henry Pohl, and Frank Hughes.
2. NASA Oral Histories interview with Arnold Aldrich.
3. Murray and Cox, *Apollo*, p. 392, 398.
4. Kraft, *Flight*, p. 336; Murray and Cox, *Apollo*, p. 432; interview with Aldo Bordano, 2016.
5. NASA Oral Histories interview with Henry Pohl.
6. Kelly, *Moon Lander*, p. 76; Murray and Cox, *Apollo*, pp. 424–425; interview with Cynthia Wells, 2016.
7. NASA Oral Histories interview with Larry Bell.
8. Interviews with Gerry Griffin and Lee Norbraten, 2016.
9. Relying particularly on Murray and Cox, *Apollo*, pp. 420–425, 440–445; Cortright, *Apollo Expeditions to the Moon*, Chapter 13.
10. NASA Oral Histories interview with Joseph "Guy" Thibodaux.
11. Interview with Marlowe Cassetti, 2017.
12. NASA Oral Histories interview with Jack Garman; Mindell, *Digital Apollo*, pp. 243–245; the hack's author was Don Eyles.
13. Tribbe, *No Requiem for the Space Age*, pp. 193–194.
14. Brooks, Crimwood, and Swenson, *Chariots for Apollo*, p. 202; Murray and Cox, *Apollo*, pp. 449–451; Compton, *Where No Man Has Gone Before*.
15. NASA Oral Histories interview with John Annexstad.
16. Mindell, *Digital Apollo*, p. 256; French and Burgess, *In the Shadow of the Moon*, p. 78.
17. NASA Oral Histories interview with Joseph Allen.
18. Ibid.
19. Scott and Jurek, *Marketing the Moon*, pp. 73–74.
20. NASA Oral Histories interview with Frank Hughes; interview with Frank Hughes, 2016.
21. De Monchaux, *Spacesuit*, p. 199.
22. NASA Oral Histories interview with Peter Armitage.
23. Interview with Caleb Fassett, 2018.

24. Beattie, *Taking Science to the Moon*, pp. 140–141; Janna Levin, "Gravitational Wave Blues," *Aeon*, April 21, 2016.

25. Interviews with Dr. Eugene Benton, 2016; Delp et al., "Apollo Lunar Astronauts Show Higher Cardiovascular Disease Mortality"; interview with Dr. Jack Miller, 2017.

26. Astronauts were Eugene Cernan, Ronald Evans, and Harrison Schmitt; Reinert, "The Blue Marble Shot"; de Monchaux, *Spacesuit*, p. 337. I have most often seen the photo credited to astronaut Eugene Cernan. Sources actually conflict on NASA's intentionality with this photo opportunity.

27. Tribbe, *No Requiem for the Space Age*, p. 181, 221; NASA Historical Data Book, Vol. 1, p. 14; interview with Marlowe Cassetti, 2017.

28. Neufeld, *Von Braun*, pp. 411, 441, 449, 451, and 452.

29. Interviews with Robert Austin and Hal Beck, respectively, 2017.

30. Interview with Wesley Ratcliff, 2017.

31. Tribbe, *No Requiem for the Space Age*, p. 10; Dunar and Waring, *Power to Explore*, p. 106.

32. Dick and Launius, *Societal Impact of Spaceflight*, pp. 58–59; Scott and Jurek, *Marketing the Moon*, pp. 75–77.

33. Oberg, *Red Star in Orbit*, pp. 100–105.

34. NASA Oral Histories interviews with Caldwell Johnson and Kenneth Young; interview with Cynthia Wells, 2016.

35. NASA Oral Histories interview with Marlowe Cassetti.

Chapter 14

1. NASA Oral Histories interview with Max Faget.

2. Interview with Guy Faget, 2017.

3. Dick and Launius, *Societal Impact of Spaceflight*, p. 430.

4. Neufeld, *Von Braun*, pp. 449, 471.

5. Neufeld, *Von Braun*, pp. 469–472; Ward, *Dr. Space*, p. 219; Neufeld, *Von Braun*, pp. 474–475.

6. Harris and *USA Today* polling data, as discussed in Tribbe, *No Requiem for the Space Age*, p. 10.

7. Lascala, Marisa. "Why Is the VMA Statue a Moonman?" *Bustle*, April 24, 2014.

8. Dick and Launius, *Societal Impact of Spaceflight*, pp. 219, 221, 224, 236.

9. Kraft, *Flight*, p. 353; NASA Oral Histories interview with Larry Bell.

10. NASA Oral Histories interview with Jack Garman.

11. Combining thoughts and arguments from NASA Oral Histories interview with Jack Garman; Dick and Launius, *Societal Impact of Spaceflight*; Tomayko, *Computers in Spaceflight*; and Mindell, *Digital Apollo*.

12. Interview with Marlowe Cassetti, 2018; NASA Oral Histories interview with Marlowe Cassetti.

13. See, for instance, William J. Broad, "Hunt is On for Scattered Blueprints of Powerful Saturn Moon Rocket," *New York Times* (May 26, 1987), p. C3.

Chapter 15

1. Dick and Launius, *Societal Impact of Spaceflight*, p. 62.

2. Dick and Launius, *Societal Impact of Spaceflight*, pp. 605–606; see the work of Dr. Marcelo Vasquez, for instance, for dosage and risk estimation; M. G. Lord, "Are We Trapped on Earth?"
3. See, for instance, Barboni et al., "Early Formation of the Moon 4.51 Billion Years Ago." *Science Advances*, 3 (Jan. 11, 2017).
4. Brooks, Crimwood, and Swenson, *Chariots for Apollo*, p. 365; Rothery, *Moons*.
5. Michael E. Newman, "To the Moon and Back . . . in 2.5 Seconds," *NIST Time Capsule* of March 26, 2018, as posted to https://www.nist.gov. For more information, see, for instance, T. W. Murphy, "Lunar Laser Ranging: The Millimeter Challenge," *Reports on Progress in Physics* 76 (June 14, 2013).
6. Rosen, "The Missing Moon Files."
7. Interviews with Marlowe Cassetti, 2017 and 2018.
8. Interview with Frank Hughes, 2018.
9. Interview with Henry Pohl, 2017.

Chapter 16

1. French and Burgess, *In the Shadow of the Moon*, pp. 81–83.
2. Interview with Gerry Griffin, 2016; NASA Oral Histories interviews with Gerry Griffin; understandably, different retellings, even from one person, have the labels varying. (In some cases, Armstrong wrote "peace" and not "talent" on one of the curves, etc.) Armstrong himself, according to Griffin, later told Griffin that he didn't recall the graphic display.
3. Ward, *Dr. Space*, p. 128.
4. NASA Oral Histories interview with Joseph "Guy" Thibodaux.
5. Interview with Gerry Griffin, 2016.
6. NASA Oral Histories interview with Peter Armitage.
7. NASA Oral Histories interview with Elmer Barton.
8. NASA Oral Histories interview with Peter Armitage.
9. Bilstein, *Stages to Saturn*, Chapter 9; Dunar and Waring, *Power to Explore*, p. 48; NASA Oral Histories interview with Aleck Bond.
10. NASA Oral Histories interview with Henry Pohl.
11. Interview with Marlowe Cassetti, 2017.
12. Interview with Aldo Bordano, 2016.
13. NASA Oral Histories interviews with Marlowe Cassetti and Henry Pohl; interview with Aldo Bordano, 2016; NASA Oral Histories interview with Caldwell Johnson.
14. NASA Oral Histories interviews with John Mayer and Peter Armitage.
15. Murray and Cox, *Apollo*, pp. 264–265.
16. NASA Oral Histories interview with Hal Beck.
17. Interviews with Henry Pohl, 2017 and 2016.
18. NASA Oral Histories interview with Thomas Moser.
19. Interview with Marlowe Cassetti, 2018; French and Burgess, *In the Shadow of the Moon*, pp. 50–51; Murray and Cox, *Apollo*, p. 448.
20. Kraft, *Flight*, p. 340; interview with Gerry Griffin, 2016.
21. NASA Oral Histories interview with Joseph Allen; meta-analysis piggy-backing on Matthew Tribbe's *No Requiem for the Space Age* and Gerard De Groot's *Dark Side of the Moon*.

BIBLIOGRAPHY

Baldwin, Ralph R. *A Fundamental Survey of the Moon.* New York: McGraw Hill, 1964.

Beattie, Donald A. *Taking Science to the Moon: Lunar Experiments and the Apollo Program.* Baltimore: Johns Hopkins University Press, 2001.

Bell, Daniel. *The Coming Post-Industrial Society: A Venture in Social Forecasting.* New York: Basic Books, 1973.

Benson, Charles D., and William Barnaby Faherty. *Moonport: A History of Apollo Launch Facilities and Operations.* NASA SP-4204, 1978.

Bilstein, Roger. *Stages to Saturn: A Technological History of the Apollo/Saturn Launch Vehicle.* Washington, DC: NASA Press, 1996.

Broad, William J. "Hunt Is on for Scattered Blueprints of Powerful Saturn Moon Rocket." *New York Times,* May 26, 1987, p. C3.

Brooks, Courtney G., James M. Crimwood, and Loyd S. Swenson, Jr. *Chariots for Apollo: A History of Manned Lunar Spacecraft.* Washington, DC: NASA, 1979.

Burrows, William E. *This New Ocean: The Story of the First Space Age.* New York: Random House, 1998.

Canup, Robin M. "Simulations of a Late Lunar-Forming Impact." *Icarus* 168 (2004), 433–456.

Cleaver, Eldridge. *Soul on Ice.* New York: Dell, 1968.

Compton, William David. *Where No Man Has Gone Before: A History of Apollo Lunar Exploration Missions.* NASA SP-4214, 1989.

Cortright, Edgar (ed). *Apollo Expeditions to the Moon: The NASA History.* New York: Dover, 2009.

DeGroot, Gerard J. *Dark Side of the Moon: The Magnificent Madness of the American Lunar Quest.* New York: NYU Press, 2006.

Delp, Michael D., Jacqueline M. Charvat, Charles L. Limoli, Ruth K. Globus, and Payal Ghosh. "Apollo Lunar Astronauts Show Higher Cardiovascular Disease Mortality: Possible Deep Space Radiation Effects on the Vascular Endothelium." *Scientific Reports* 6, No. 29901 (2016).

De Monchaux, Nicholas. *Spacesuit: Fashioning Apollo.* Cambridge, MA: MIT Press, 2011.

Dethloff, Henry C. *Suddenly Tomorrow Came: A History of the Johnson Space Center.* NASA SP-4307, 1993.

Dick, Steven J. (ed.). *NASA's First 50 Years: Historical Perspectives.* NASA SP-2010-4704, 2010.

Dick, Steven J., Stephen J. Garber, and Jane H. Odom (compilers). *Research in NASA History: A Guide to the NASA History Program.* NASA SP-4543, 2009.

Dick, Steven J., and Roger D. Launius. *Societal Impact of Spaceflight.* NASA SP-2007-4801, 2007.

Dickson, Paul. *Sputnik: The Shock of the Century.* New York: Walker, 2001.

Dunar, Andrew J., and Stephen P. Waring. *Power to Explore: A History of Marshall Space Flight Center 1960–1990.* NASA SP-4313. Washington, DC: NASA Publications, 1999.

Ezell, Linda Neuman. *NASA Historical Data Book, Volume II: Programs and Projects 1958–1968*. NASA SP-4012. Washington, DC: NASA Publications, 1988.

Farber, David (ed). *The Sixties: From Memory to History*. Chapel Hill: University of North Carolina Press, 1994.

French, Francis, and Colin Burgess. *In the Shadow of the Moon: A Challenging Journey to Tranquility, 1965–1969*. Lincoln, NE: University of Nebraska Press, 2007.

Fries, Sylvia D. NASA Engineers and the Age of Apollo. NASA SP-4104, 1992.

Gainor, Chris. *To a Distant Day: The Rocket Pioneers*. Lincoln: University of Nebraska Press, 2008.

Gitlin, Todd. *The Sixties: Years of Hope, Days of Rage*. New York: Bantam, 1989.

Godwin, Robert (ed.). *Apollo 17: The NASA Mission Reports Volume One*. Burlington, Canada: Apogee, 2002.

Grinter, Kay. "Chosen Few Apollo Trailblazers Wore High Heels." *Spaceport News*, March 2009, 7–8.

Harford, James. *Korolev: How One Man Masterminded the Soviet Drive to Beat America to the Moon*. New York: Wiley & Sons, 1997.

Hickman, Homer. *Rocket Boys*. New York: Dell Publishing, 1998.

Holt, Nathalia. *Rise of the Rocket Girls: The Women Who Propelled Us, from Missiles to the Moon to Mars*. New York: Little, Brown, 2016.

Ifrah, Georges. *A Universal History of Computing: From the Abacus to the Quantum Computer*. New York: Wiley, 2001.

Jacobsen, Annie. *Operation Paperclip: The Secret Intelligence Program that Brought NAZI Scientists to America*. New York: Little, Brown, 2014.

Jastrow, Robert, and Homer E. Newell. "Why Land on the Moon?" *Atlantic Monthly* 211, No. 2 (August, 1963), 41–45.

Jules-Verne, Jean. *Jules Verne: A Biography*. New York: Taplinger Publishing, 1976.

Junior League of Huntsville Alabama. *Huntsville Heritage Cookbook*. Huntsville: Hicklin Printing Co., 1967.

Kelly, Thomas J. *Moon Lander: How We Developed the Apollo Lunar Module*. Washington, DC: Smithsonian Institution Press, 2001.

Kemeny, John G. *Man and the Computer*. New York: Charles Scribner's Sons, 1972.

Kennan, Erlend A., and Edmund H. Harvey, Jr. *Mission to the Moon: A Critical Examination of NASA and the Space Program*. New York: Morrow, 1969.

Kennedy, John Fitzgerald, Deirdre Henderson, and Hugh Siley. *Prelude to Leadership: The European Diary of John F. Kennedy: Summer 1945*. Washington, DC: Regnery, 1995.

Kluger, Jeffrey. *Apollo 8: The Thrilling Story of the First Mission to the Moon*. New York: Henry Holt, 2017.

Kraft, Chris. *Flight: My Life in Mission Control*. New York: Dutton, 2001.

Kranz, Gene. *Failure Is Not an Option: Mission Control from Mercury to Apollo 13 and Beyond*. New York: Simon and Schuster, 2009.

Kurlansky, Mark. *1968: The Year That Rocked the World*. New York: Random House, 2004.

Lasby, Clarence G. *Project Paperclip: German Scientists and the Cold War*. New York: Antheneum, 1971.

Launius, Roger. *NASA: A History of the U.S. Civil Space Program* (2nd ed.). Malabar, FL: Krieger, 2001.

Launius, Roger. *Spaceflight and the Myth of Presidential Leadership*. Champaign: University of Illinois Press, 1997.

Launius, Roger, and Dennis R. Jenkins. *Coming Home: Reentry and Recovery from Space.* Washington, DC: NASA, 2011.

Launius, Roger, John M. Logsdon, and Robert W. Smith. *Reconsidering Sputnik: Forty Years Since the Soviet Satellite.* Cornwall: Harwood Academic, 2000.

Logsdon, John M. *The Decision to Go to the Moon: Project Apollo and the National Interest.* Cambridge, MA: MIT Press, 1970.

Lord, M. G. "Are We Trapped on Earth?" *Discover Magazine,* June 2006.

McDougall, Walter A. . . . *the Heavens and the Earth: A Political History of the Space Age.* New York: Basic Books, 1985.

Mindell, David. *Digital Apollo: Human and Machine in Spaceflight.* Cambridge, MA: MIT Press, 2008.

Mooncat Publications. *The Manned Spaceflight Patents of Max Faget.* San Diego: Mooncat Publications, 2014.

Mudgway, Douglas J. *Uplink-Downlink: A History of the Deep Space Network, 1957–1997.* NASA SP-20014227, 2001.

Murray, Charles, and Catherine Bly Cox. *Apollo: The Race to the Moon.* New York: Touchstone, 1989.

National Aeronautics and Space Administration (NASA). *Johnson Space Center Oral History Project: NASA at 50.* Johnson Space Center History Office, 2010. (See: http://www.jsc.nasa.gov/history/oral_histories/oral_histories.htm)

National Aeronautics and Space Administration (NASA), Manned Spacecraft Center. *Results of the First United States Manned Orbital Space Flight.* Washington, DC: U.S. Government Printing Office, 1962.

National Aeronautics and Space Administration (NASA). *Saturn V Flight Manual, SA 503.* Washington, DC: U.S. Government Printing Office, November 1, 1968.

Neufeld, Michael J. *Von Braun: Dreamer of Space, Engineer of War.* New York: Vintage, 2007.

Newport, Curt. *Lost Spacecraft: The Search for Liberty Bell 7.* Ontario: Apogee, 2002.

Oberg, James. *Red Star in Orbit.* Random House: New York, 1981.

Pfeiffer, John. "Moon Mysteries Ranger Didn't Solve." *Popular Mechanics,* November 1964.

Piszkiewicz, Dennis. *Wernher von Braun: The Man Who Sold the Moon.* London: Praeger Publishers, 1998.

Rabinowitch, Eugene, and Richard S. Lewis (eds.). *Man on the Moon: The Impact on Science, Technology, and International Cooperation.* New York: Basic Books, 1969.

Reeves, Robert. *The Superpower Space Race: An Explosive Rivalry through the Solar System.* New York: Plenum, 1994.

Reinert, Al. "The Blue Marble Shot: Our First Complete Photograph of Earth." *Atlantic Monthly,* April 12, 2011.

Rosen, Julia. "The Missing Moon Files." *Discover Magazine,* November, 2016, 68–72.

Rothery, David A. *Moons: A Very Short Introduction.* New York: Oxford University Press, 2015.

Scott, David M., and Richard Jurek. *Marketing the Moon: The Selling of the Apollo Lunar Program.* Cambridge, MA: MIT Press, 2014.

Seamans, Robert C., Jr. *Project Apollo: The Tough Decisions.* Monograph in Aerospace History No. 37. NASA SP-2005-4537, 2005.

Shetterly, Margot Lee. *Hidden Figures: The American Dream and the Untold Story of the Black Women Mathematicians Who Helped Win the Space Race.* New York: William Morrow, 2016.

Steigerwald, David. *The Sixties and the End of Modern America.* New York: St. Martin's Press, 1995.

Stephenson, F. R., L. V. Morrison, and C. Y. Hohenkerk. "Measurement of the Earth's Rotation: 720 BC to AD 2015." *Proceedings of the Royal Society A* 472, 20160404 (2016).

Swedin, Eric G., and David L. Ferro. *Computers: The Life Story of a Technology.* Westport, CT: Greenwood, 2005.

Tomayko, James. *Computers in Spaceflight: The NASA Experience.* Washington, DC: NASA, 1988.

Tribbe, Matthew D. *No Requiem for the Space Age: The Apollo Moon Landings and American Culture.* New York: Oxford University Press, 2014.

Tsiao, Sunny. *"Read You Loud and Clear!" The Story of NASA's Spaceflight Tracking and Data Network.* NASA SP-2007-4233, 2008.

Turnill, Reginald. *The Moonlandings: An Eyewitness Account.* New York: Cambridge University Press, 2003.

U.S. Government, Navy Department, Office of the Chief of Naval Operations. *Dictionary of American Naval Fighting Ships, Vol. 3.* Washington, DC: U.S. Government Printing Office, 1968.

Van Nimmen, Jane, and Leonard C. Bruno, with Robert L. Rosholt. *NASA Historical Data Book, Volume I: NASA Resources, 1958–1968.* NASA SP-4012, 1976.

Verbeek, J. R. *"V2-Vergeltung" from The Hague and Its Environs: Deployment of the V2 Rockets and the Terrors for the City and Her Inhabitants.* Den Haag: Almere, 2005.

Von Braun, Wernher. "Crossing the Last Frontier." *Collier's,* March 22, 1952.

Von Braun, Wernher. "The Journey." *Collier's,* October 18, 1952.

Von Braun, Wernher, and Frederick J. Ordway. *History of Rocketry and Space Travel.* New York: Thomas Y. Crowell Co., 1975.

Von Braun, Wernher, and Cornelius Ryan. "Can We Get to Mars?" *Collier's,* April 30, 1954.

Von Ehrenfried, Manfred. *The Birth of NASA: The Work of the Space Task Group, America's First True Space Pioneers.* New York: Springer-Praxis, 2016.

Ward, Bob. *Dr. Space: The Life of Wernher von Braun.* New York: Naval Institute Press, 2013.

Ward, Peter D., and Donald Brownlee. *Rare Earth: Why Complex Life Is Uncommon in the Universe.* New York: Copernicus, 2000.

Watkins, Billy. *Apollo Moon Missions: The Unsung Heroes.* Lincoln: University of Nebraska Press, 2007.

Williams, Mike. "Moon Dust Mystery Mostly Solved." *Rice News,* November, 2013.

Wolfe, Tom. *The Right Stuff.* New York: Farrar, Straus, and Giroux, 1979.

INDEX

Figures are indicated by an italic *f* following the page/paragraph number

Aaron, John, 191–92, 199, 200
Abernathy, Ralph, 182
ablator, 34–35
accelerometer, 29–30
Agnew, Spiro, 195
Aldridge, Arnold, 56–57, 199
Aldrin, Buzz, 188
Allen, Harvey, 34
Allen, Joe, 206–7, 239
Altamont Pass, 195
American astronauts. *See also* Russian
    cosmonauts
  Apollo 8 astronauts as *Time* magazine
      "man of the year," 168
  astronaut safety, 91
  celebrity status of, 65–66
  deploying lunar science projects, 206*f*
  as field geologists, 206–7
  first Earth orbit, 64–66
  frustrated relationship with science, 206
  loss of, 123–24
  navigating home by star reference, 123–24
  parade for, 66
  radiation dosage on, 189
  reaction to Saturn V's launch, 159
  space simulations for, 121–23
  space walk, 104
  training female, 81–82
  training for lunar landing, 151–52
  training of, 121–23
  in vacuum chamber, 120
amp trap, 192
Anders, William, 166, 168

Apollo 8
  astronauts jarring takeoff, 159
  barbecue mode, 161–62
  Christmas Eve broadcast, 166
  command mistype, 167
  communication time delays, 163–64
  earthrise photograph, 166
  on far side of the Moon, 164–65
  Gerry Griffin on, 159–60
  launch of, 159–60
  lunar orbit, 164–66
  outbound trip, 159–62
  path of, 161
  radio black-out, 164
  re-entry violence, 167–68
  return to Earth, 166–68
  space sickness during, 160–61
  temperature extremes, 161–62
Apollo 11
  Armstrong's famous statement
      during, 186
  astronaut quarantine, 189–90
  celebrations for, 189
  communication systems, 186–87
  empty fuel tank warning, 184
  lander descent, 182–84
  landing radar, 183
  launch of, 182
  lunar landing module setting down, 185
  Moon liftoff, 188
  protests on, 182
  radiation dosage on astronauts, 189
  return to Earth, 188–89

Apollo 11 (*cont.*)
  splash down, 189
  stepping on the Moon, 186–87
  television audience and, 186
  trip to the Moon, 182
  TV signal, 186–87
  United States flag and, 181–82, 188
  warning alarms during, 183–84
Apollo 13
  astronauts dehydrating during, 202
  budgeting electricity, 200
  carbon dioxide filters, 201
  Cassetti on near tragedy of, 197
  entering barbecue mode, 202
  inter-center tensions discussing, 203
  liquid oxygen tank malfunction, 197–98
  near tragedy of, 197
  oxygen tank explosion, 198–99
  preparing for, 197
  preparing for re-entry, 202
  stirring oxygen tank, 198–99
Apollo 13 (film), 203–4
Apollo applications, 129–30
Apollo capsule, 35, 133, 150, 213
Apollo display, 197
Apollo missions. *See also* Apollo 11; Apollo
    13; Apollo 8; lunar landing module;
    return-on-investment
  American's interest in, 211–12
  Apollo 1, 135–36, 137–38
  Apollo 7, launching of, 157–58
  Apollo 9, 161, 173–74, 175–76
  Apollo 10, 176–79
  Apollo 12, 191–95
  Apollo 14, 204–5
  Apollo 15, 142*f*, 207
  Apollo 17, 210
  autopilot for, 116
  canceled, 220–21
  close of, 210
  command module, 133–34, 138
  declining television ratings, 211–12
  extra-sensory perception
    experiments, 205
  first manned, 157–58
  geology field and, 205
  goal of, 47–48, 59

  modules, 134*f*
  negativity facing, 80
  postpartum effects, 211
  science experiments, 205, 226–27
  service module, 133–34
  simulated countdown, 135–36
  spacesuit for (*see* spacesuits)
  step-by-step illustration, 178*f*
Apollo Soyuz program, 212–13
Apple Computer company, 219
Armitage, Peter, 44–45, 232–33, 235–36
Armstrong, Neil, 124, 151–52, 184–85, 186,
    188, 230–31
Arrow jet fighter, 37
ascent engine, 89–90
Astronauts. *See* American astronauts;
    Russian cosmonauts
Atlas missile, 39–40, 45, 47
Austin, Bob, 18–19, 75, 101–2, 210
automated systems, 90–91

Bales, Steve, 13–14, 183–84
barbecue mode, 161–62, 202
Barton, Elmer, 233
Bassett, Charles, 123–24
Bay of Pigs invasion, 46
Beck, Hal
  on Houston Space Center, 52
  on lunar landing sites, 81
  on Moon's gravitational pull, 165–66
  on Moon trajectories, 70–72
  on NASA bureaucracy, 236–37
  photograph of, 73*f*
  Schmitt and, 211
  on Space Task Group, 32–33
Bell, Larry, 103–4, 136, 219
Bell Aerospace, 119–20
belly band, 40
Benton, Eugene, 189
blockhouse, 100
"Blue Marble" photograph, 210
Bond, Aleck, 34–35, 52
Bordano, Aldo, 93, 94, 199–200, 234
Borman, Frank, 160, 166
Bourgin, Simon, 166
Bradbury, Ray, 12
Brezhnev, Leonid, 175

broom test, 75
Brown, Robert
  in advisory role, 190, 221
  on barbecue mode, 161–62
  on calculating consumables, 139
  early life of, 9–10
  on guidance instruments, 29–30
  hiring new employees, 93
  on John Norton, 115
  joining Mission Planning and Analysis
    division, 93
  on ascent from Moon, 89
  lunar landing navigation assignment, 90–
    91
  mementos of, 229
  on Moon trajectories, 70
  photograph of, 228f
  in retirement, 229
  running failure analysis project, 221
  on Surveyor program, 70, 90
  on teamwork atmosphere, 236
Bush, George H. W., 224–25
Bush, George W., 224–25

Caltech gathering, 230
Canada aerospace, 37
Cape Canaveral, 53, 54, 57, 82
Capsules
  early model of, 67f
  returning, 64–65
Carruth, Ralph, 101f
Cassetti, Marlowe
  on Apollo 13 near tragedy, 203
  on Apollo postpartum, 214
  calculating Apollo's total weight, 154
  on close of Apollo, 210
  on communication relay, 57
  competing for computer time, 72
  on control thruster problems, 65
  coordinating early launches, 24
  early life of, 7–8
  on engineers' obsessive dedication, 185
  experiencing orbit of globe, 58
  at government's aeronautical labs, 21
  guiding Gemini launches, 105
  on Houston Space Center, 52
  on Huntsville, Alabama, 24–25

  on hydrogen fueled rockets, 72
  on lander's weight, 139–40
  at Langley, Virginia, 21
  on early launch films, 36–37
  on model airplanes, 8–9
  on NASA bureaucracy, 237–38
  on NASA's start, 25
  on personal computers, 220
  photograph of, 9f
  predicting future Moon missions, 222
  on racial segregation, 96
  reacting to Kennedy's speech, 48
  in retirement, 227
  retiring from NASA, 220
  on sight selection for Manned Spaceflight
    Center, 50–51
  on Skylab program, 214
  on teamwork atmosphere, 234–35
  underestimating von Braun's engineering
    skills, 234
  on von Braun's organization
    structure, 49–50
Cavett, Dick, 216–17
CBS News, 180
Cernan, Eugene, 126–27, 179
Chaffee, Roger, 136
Christmas Eve broadcast, 166
circumlunar voyage, support for, 155–57
Civil rights, 96–97
Civil Rights Act, 96, 109
Cleaver, Eldridge, 147
Cold War paranoia, 47
combustion instability, 79, 84–86. See also
    pogo effect
communications satellites, 83
communication system, global,
    55–56, 163–64
computer chip, 72
computer programming, 116–17
computers
  electronic, 32–33
  female technical secretaries as first, 31–32
  integrated circuits, 72
  lunar trajectory computations, 70–71
  machine time for computations, 70–72
  memory, 118
  MIT's delivery of, 116–17

computers (*cont.*)
  personal computers, 219–20
  programmable computer, 118
  punch cards, 70–72
  rope mothers, 118
  semiconductor chips, 220
  software, 115–16
  storing programs on spacecraft, 117–18
  teletype, communication via, 140
computing time, 140
Congressional Space Act of 1958, 25
Conrad, Charles, Jr., 193, 194*f*
control thrusters, 64–65, 99, 125–26, 200
Cooper, Gordon, 237–38
Copernicus Crater, 131*f*
Corona satellite, 64
cosmic rays, 188–89, 225. *See also* space
  radiation
countdown, simulated, 135–36
crawlers, 76, 114, 222
Cronkite, Walter, 133, 144, 180, 187

direct ascent debates, 42–43
Disney, Walt, 13*f*
dispersion analysis, 62–63, 176–77, 178*f*
Dunbar's number, 231
dynamic test stand, 100

Earth
  gravitational pull of, 161
  Moon's gravitational effect on, 60
  phase-locking to the Moon, 60
  relative scale of Moon and, 61*f*
Earth-centered to Moon-centered
  coordination, 162
Earth-Moon distance, 226
Earth orbit rendezvous, 43–44
Earth's atmosphere, 43
Eisenhower, Dwight, 17–18, 25, 43–44
electrocardiogram (EKG) monitoring, 219
electromagnetic waves, 209
electronic computers, 32–33
engineers
  civil rights and, 130
  families of, 96
  female, 108–9
  minority, 109, 211

NASA's trust in, 232–33
  obsessive dedication of, 185
  responsibility of, 233
  on teamwork atmosphere, 236
environmental chamber, 104–5, 120, 138
equal opportunity/affirmative action, 216
escape tower, 37, 38*f*, 47
Explorer 1, 24
extra-sensory perception experiments, 205

F-1 engine, 48–49, 84, 100
Faget, Ann, 95–96
Faget, Carol, 95–96
Faget, Guy Henry, 5, 170, 172, 215
Faget, Maxime "Max"
  designing escape tower, 37
  designing hypersonic aircraft, 21–22
  on flying to the Moon, 42
  founding Space Industries, 215
  on John Glenn's Earth orbit, 66
  Johnson and, 235
  at Langley, Virginia, 21–22
  on lunar lander plan, 87–88
  on lunar orbit rendezvous, 43–44
  on Moon disease risk, 95
  on oxygen problems, 133
  personality of, 22
  photograph of, 5*f*, 67*f*, 171*f*
  quirky brilliance of, 6, 95
  recognizing importance of
    management, 237
  retiring from NASA, 215
  on shuttle idea, 170
  spacecraft designer role, 33–36
  spacecraft testing, 94–95
  on Space Shuttle program, 215
  submarine life of, 6
  von Braun and, 172
  on Apollo TV signal, 186–87
  during WWII, 4–6
Fairchild Semiconductor, 72
Fassett, Caleb, 208–9
female astronauts and cosmonauts,
  81–82, 210
female computers, 31–32, 33
female engineers, 108–9
"Flash Gordon" (serial), 10

flying bedstead, 151–52, 152*f*
food safety, 218–19
fuel cells, 107, 113–14, 200
fuel tanks, 75–76, 125–26

Gagarin, Yuri, 45
Garman, Jack
  on Apollo 8, 162
  on Apollo 11 alarms, 183
  on Apollo 14, 204
  on Apollo guidance computer, 219–20
  on control center's upgrade, 204
  joining NASA, 115–16
  preparing "cheat sheets," 181
Gemini missions, 59, 98, 105–6, 107–8,
  124, 134*f*
General Electric, 131
Gilruth, Robert, 21, 31–32, 39–40, 46–47,
  50–51, 53, 234, 237
gimbal lock, 179
Glenn, John, 64–66
global communication system, 55–56,
  112, 163–64
Goddard, Robert, 3–4, 9
Goddard Space Flight Center, 56, 57
gravimeter, 209
gravitational waves, 209
Griffin, Gerry
  on Apollo 8, 159–60
  on Apollo 12, 192
  at Caltech gathering, 230–31
  on NASA bureaucracy, 238
  on NASA culture, 232
  on radio black-out, 165
Grissom, Gus, 135, 136
Grumman Company, 119–20, 138–39, 190
guidance instruments, 30, 191–92
gyroscope, 29–30, 141–42

Ham (chimpanzee), 44–45
Hamilton Standard, 91–92
Harris, Ruth Bates, 216
Haukohl, Guenther, 15, 53, 236–37
Heimburg, Karl, 78
Henderson, Mac, 156, 191
Houbolt, John, 67–68
Houston, test facilities in, 77–78

Howard Hughes's aircraft company, 70
Hubble Space Telescope, 106
Hughes, Frank
  Anders and, 168
  on Apollo 1 tragedy, 136
  on communication time delays, 163–64
  creating lifelike simulations, 122–24
  joining NASA, 120–21
  on lunar landing set down, 185
  on Moon landing skeptics, 207–8
  on rare day off, 157
  in retirement, 227–28
  on space sickness, 161
Humble Oil Company, 51
Huntsville
  blossoming of, 24
  celebration of Moon announcement, 48
  celebration of Moon landing, 189
  German relocation to, 12
  growth of, 129
  as "Rocket City," 24, 224
  sonic impact on, 26–27
  symphony orchestra of, 24
  Thanksgiving in, 145–46
  today, 224
hydrogen-burning engines, 72–75, 76
hypergolic fuels, 89–90, 98
hypersonic aircraft, designing, 21–22

ice crystals, 123–24, 157–58
inertial guidance, 29–30
instrument unit, 141–42, 143*f*, 149
International Business Machines
  (IBM), 219
International Latex Corporation, 92,
  104, 208

Jet Propulsion Laboratory, 42, 70, 119
Johnson, Caldwell, 48, 52–53, 62, 105, 133,
  212–13, 235, 238
Johnson, Katherine, 33
Johnson, Lyndon Baines
  condemning urban violence, 148
  on female astronauts, 81–82
  on NASA's budget, 96, 129–30
  NASA support of, 84
  recommending Moon landing, 46–47

Johnson, Lyndon Baines (*cont.*)
 on reelection, 149
 signing Civil Rights Act, 96
 on Sputnik, 17
 sworn in as president, 83
 as technocracy disciple, 44
Johnson Space Center, 51, 215–16,
 218, 220–21
joint space exploration, 82, 212–13
Jupiter rocket, 14, 19

Kelly, Thomas, 87–88, 89, 119–20,
 185–86, 188
Kennedy, John F.
 on Apollo project, 44
 assassination of, 83
 committing to Moon landing, 47–48
 health problems of, 45–46, 82
 at Marshall Space Flight Center, 82*f*
 okaying Bay of Pigs invasion, 46
 on Soviet joint space exploration, 82
 on space race, 31
 on superweapons, 4–5
 and technocracy, 44, 64
Kennedy, Joseph, Jr., 4–5
Khrushchev, Nikita, 16, 17, 20, 81, 83
King, Martin Luther, Jr., 77, 130, 147, 151
Komarov, Vladimir, 137
Korolev, Sergei Pavlovich
 death of, 118
 designing first artificial satellite, 17
 developing the Voskhod, 83
 joining Soviet Union rocket program,
  15–17
 on Sputnik II, 20
Kraft, Christopher
 on Apollo 13 near tragedy, 199
 coordinating capsule recovery with
  Navy, 158–59
 on Gilruth, 31
 investigating Apollo 1 tragedy, 137–38
 on John Glenn's Earth orbit, 64
 as minor celebrity, 108
 Mission Control and, 59
 on NASA bureaucracy, 238
 on Saturn V maiden voyage, 144
 on scientists vs. engineers, 80–81

on Titan rocket launch, 113
von Braun and, 53

Laitin, Joseph, 166
Lang, Fritz, 9
launch days, 112
launch pad, design of, 113–14
Leonov, Alexei, 103–4
lifeboat study, 201
lightning, 191–92
Lindbergh, Charles, 175
liquid oxygen, 3–4, 120–21
long-duration space missions, 225
Low, George, 155–56, 216
lunar-centered coordinate system, 162
Lunar Lander Research Vehicle, 152*f, See
 also* flying bedstead
lunar landing, sites for, 81
lunar landing module
 ascent engine, 89–90
 cracked parts in, 139
 docking command module to, 173–74
 first generation, 87–89
 full-scale model, 91
 functioning as a lifeboat, 200–1
 photograph of, 88*f*
 prototype for, 119–20
 size and weight of, 173
 support for, 87
 testing of, 138–39
 weight of, 139–40, 153–54
 wiring issues, 139, 149
Lunar orbit rendezvous, 43–44, 67–69
lunar receiving laboratory, 95, 208–9
lunar television signal, 207
Luna series, 24, 63, 118–19, 151
Lunney, Glynn, 151
Lyndon B. Johnson Space Center. *See*
 Manned Spaceflight Center

Manned Spaceflight Center, 50–53,
 93, 147–48
Mars, 55, 68, 163–64, 172, 189, 209,
 214, 224–25
Marshall Space Flight Center, 26,
 86, 222–24
Max Q, 39

Mayer, John, 235–36
Maynard, Owen, 37, 39
McQuarrie, Ralph, 180
Mercury capsules, 65
Mercury missions, 33, 54, 59, 134*f*
Merritt Island launch facility, 53–55
minority engineers, 108–9
Mission Control
   during Apollo 11, 185
   during Apollo 13, 202
   astronauts communicating with, 163
   audio lines with capsule and, 135
   computer failure during Apollo 8, 162–63
   computer programs for, 93
   computers running, 162–63
   design of, 53, 59
   global stations communicating with, 56–57
   photograph of, 58*f*
   running simulations, 181
Mission Planning and Analysis division, 93–94, 157, 214
Mississippi test facility, 127
Mitchell, Edgar, 205
MIT's Instrumentation Laboratory, 72, 115–17, 181, 204–5
mobile launch concept, 55, 141, 176, 222
Moon
   features of, 62
   gravitational effect on Earth, 60
   mass concentrations in, 132
   measuring distance to, 226
   NASA plotting orbits of, 62–63
   orbit of, 60
   origin of, 225–26
   phase-locking of, 60
   placing United States flag on, 181–82
   relative scale of Earth and, 61*f*
   science experiments, 226–27
   scientific discoveries of, 225
   scientific priorities for, 80–81
   shooting probes to, 70
"Moon Day" idea, 195
Moon landing
   American public's declining interest in, 195
   Americans' disbelief in, 193–95

earth orbit rendezvous, 43–44
   as fake, 193–95
   Johnson's recommendation to Kennedy on, 46–47
   lunar orbit rendezvous, 43–44, 67–69
   media coverage of, 180
   NASA readying first, 179
   one-ship approach, 68
   world's enthusiasm for, 197
moonquakes, 62, 205, 217
Moon rocks, 90, 92, 188, 190, 197, 208–9, 225–26
Moon virus, 95
Moser, Tom, 69, 109–10, 181–82, 237
MTV (Music Television), 217–18
Mueller, George, 237

National Academy of Sciences, 95
National Aeronautics and Space Administration (NASA). *See also* Johnson Space Center; Manned Spaceflight Center; Marshall Space Flight Center; Mission Control; return-on-investment
   Alabama vs., 96–97
   brainstorm city planning, 64
   building launch facility, 53–55
   bureaucracy of, 236–38
   buying computers, 72
   collaborative spaceflights with Russia, 212–13
   developing global communication system, 55–56
   diverse workforce of, 108, 216
   eliminating final three missions, 197
   emergence of, 25
   employment record, 216
   environmental chamber, 104–5, 120, 138
   federal budget for, 96, 129–30, 190, 217
   hard-wired data link between facilities, 57
   housing options near, 52, 93–94, 95–96
   inter-center tensions, 203
   James Webb as head of, 44
   Jet Propulsion Laboratory, 42, 70, 119
   Johnson's support of, 84
   legislative oversight, 216
   lunar television signal, 207

National Aeronautics and Space
 Administration (NASA) (*cont.*)
 management mistakes, 237
 negativity facing, 80
 opening Manned Spaceflight Center,
  50–53
 organizational culture, 231–33
 progress of, 31
 space emergencies, 123–24
 success for, 152–53
 teamwork of, 234–36
Night of the Living Dead (film), 148
Nixon, Richard, 188, 190
Norbraten, Lee, 93, 156–57
Norton, John, 115
Nova rocket, 42–43

Oberth, Hermann, 3–4, 9
Operation Aphrodite, 4
Operation Breakthrough, 175
orbital paths, 43, 57
Osgood, Cathy, 32, 33, 111
Outer Space Treaty of 1967, 181–82

pad test, 192
Parnell, Tom, 50–51, 193–94
Patched Conic Technique, 70
Pegasus satellite, 102*f*, 106–7
Petrone, Rocco, 217
pitch rotation, 69
pogo effect, 74, 107, 145, 149, 152–53, 177.
  See also combustion instability
Pohl, Henry
 accumulating NASA responsibility, 27
 on Apollo 13 near tragedy, 200
 on Apollo 1 tragedy, 138
 on Apollo spacecraft interior, 130–31
 childhood of, 6–7
 considers return to Huntsville, 79
 control thrusters and, 99, 125–26, 200
 on criteria for successful launch, 37
 developing model rockets, 27–28, 29*f*, 78
 as director of engineering, 215–16
 early career of, 14–15
 on Faget feedback, 95
 first encounter with von Braun, 50
 in Houston, 77–79

joining Propulsion and Power group, 79
 joining Test Lab, 15
 as movie producer, 28
 on NASA bureaucracy, 236–37
 on NASA culture, 233
 on NASA leaders, 234
 on one-page work summaries, 86
 overseeing explosives on Apollo, 155–56
 photograph of, 228*f*
 reacting to Kennedy's challenge, 48
 on repeating Apollo program, 238
 in retirement, 228–29
 on Saturn test, 26
 on teamwork atmosphere, 235
 on Thibodaux, 79
 on von Braun, 49
pressurized space flight, 102–3
probe. *See* unmanned probe
programmable computer. *See* computers
Project Horizon, 24
Project Overcast, 1–2
Project Paperclip, 2

R-7 rocket, 16–17
race, mixing of in NASA's early years, 32–33
racial equality, 130
racial segregation, 77, 96–97
radar systems, 91
radiation. *See* space radiation
Ranger 3 probe, 70
Ranger 4 probe, 70
Ratcliff, Wesley, 109, 193–94, 211
receiving laboratory, 189–90
Redstone Arsenal's ballistic missile
  program, 12, 15
re-entry, heating of, 34–35
Rees, Eberhard, 217
rendezvous at the Moon. *See also* earth orbit
  rendezvous; lunar orbit rendezvous
 description of, 110–12
 first attempt for, 112–14
return-on-investment
 computer development, 219
 Earth-orbiting satellites, 218
 medical technology, 219
 semiconductor chips, 220
 upgraded food safety, 218–19

Rice University, 51

Rocketdyne company, 74

rockets
    Army vs. Navy, 19–20
    fuel, 40–41, 49, 74, 85, 144, 152–53, 176
    and gravity, 20
    movement of, 55, 76
    stages during flight, 20
    worldwide fascination of, 9

roll rotation, 69

rovers, 207–8

Russia. *See* Soviet Union

Russian cosmonauts. *See also* American
    astronauts; Soviet rocket program
    loss of, 137
    Nikita Khrushchev and, 81

Salyut I, 211–12

Saturn rockets
    combustion instability, 79, 84–86
    destroying themselves, 27–28
    direct ascent debates about, 42–43
    launch pad design, 113–14
    scale models of, 27–28
    static test phase, 26–27

Saturn V, first stage
    first launch of, 145
    fuel for, 72–74
    pogo effect, 149
    static test firing, 102*f*
    testing of, 99–102

Saturn V, second stage
    combining oxygen with hydrogen, 74–75
    first launch of, 145
    testing of, 102, 127
    trimming weight from, 127–28

Saturn V, third stage
    testing of, 140–41
    weld-line problems, 140–41

Saturn V rocket
    aerial view of fully stacked, 142*f*
    "all up" test, 144–45
    combustion instability, 74
    dynamic test stand for, 100
    engine cracks, 86
    engine failures, 153
    first launch of, 141

    instrument unit, 143*f*
    maiden voyage of, 144–45
    numerics of, 150
    photograph of, 134*f*
    range safety decision, 149
    second launch of, 149–51
    sense of balance, 141–42
    simulated countdown, 135–36
    testing of, 99
    transporting, 128
    weight of, 127–28

Schmitt, Harrison, 206–7, 211

Schweickart, Russell "Rusty," 161

Scott, David, 124

See, Elliot, 123–24

segregation. *See* racial segregation

Shea, Joe, 131, 135, 137

Shepard, Alan, 47

simulated countdown, 135–36

simulations. *See* space simulations

six degrees of freedom, 69

Skylab program, 214

solar flares, 167, 181, 225

solar wind, 205

sonic effects, 26–27

sonic measurements, 27

soviet rocket program. *See also* Luna series;
    Russian cosmonauts
    accomplishments of, 30–31
    catastrophes of, 30
    embracing automated systems, 91
    launching first human into space, 45
    launching Vostok craft, 45
    Salyut I space station, 211–12
    successful docking mission, 174–75
    successful unmanned missions around
        the Moon, 157, 159
    Zond spacecraft, 133, 157, 159

Soviet Union
    condolences on Apollo 1 tragedy, 137
    conscripting German rocket
        scientists, 15–16
    joint space exploration, 82
    Khrushchev discovering rocket
        program, 16
    Khrushchev removed from power, 83
    launching first artificial satellite, 16–17

Soyuz capsule, 137
"Space, Science, and Urban Life"
    conference, 64
spacecraft
    air inside of, 102–3
    design of, 33–36
    testing of, 99–102
Space Exploration Initiative, 224–25
Space Launch System, 222, 224
Space Policy Directive, 224
space race, 22, 30
    accelerating, 54–55
    challenges of, 25–26
    end of, 133, 159–60, 168
    engineers wanting to join, 236
space radiation, 25–26, 189, 209, 225. *See
    also* cosmic rays
Space Shuttle program, 215, 217
space sickness, 160–61, 176
space simulations
    Apollo simulators, 121–23
    astronauts in, 120
    field of, 120–23
    star ball, 122–23
space station, 49, 175, 212, 220, 225
spacesuits, 91–92, 104, 126–27, 136, 208
Space Task Group, 31–33, 52
space travel, long-duration, 209, 225
space walk, 103–4, 126
speed, absolute versus relative, 177
Sputnik, 16–18, 19
Sputnik II, 20
spy satellites, 63, 64
Stafford, Thomas, 127
star ball, 122*f*, 122–23
static test firing, 15, 102*f*
static test stand, 15, 100, 101*f*, 222–23
Student Nonviolent Coordinating
    Committee, 96, 130
Stuhlinger, Ernst, 25–26, 106–7, 186
Super Guppy, 128, 129*f*
surveyor program, 70, 90–91, 119, 194*f*
survival couch, 35–36, 36*f*

teletype machine, 56–57, 140, 154
temperature extremes in space,
    25–26, 161–62

Tereshkova, Valentina, 81
Thibodaux, Guy, 21, 22, 74, 79, 203, 232
Thomson, Jerry, 79
thruster rockets. *See* control thrusters
Time magazine, 168–69
titanium tanks, testing of, 125–26
Titan rocket, 98
trajectories to the Moon, 69–70
"Trip around the Moon" (television
    program), 13–14
Tsiolkovsky, Konstantin, 3–4
turbopumps, 26
2001: A Space Odyssey (film), 148

University of Houston, 71, 99, 140
unmanned probe
    entering lunar orbit, 131–32
    landing on lunar surface, 118–19
unmanned test, 40–41
Urey, Harold, 81
U.S. Centers for Disease Control, 219
USS Guavina, 4–6
U.S. Space and Rocket Center, 224
USSR. *See* Soviet Union

V-2 weapon, 3*f*, 3–4
vacuum chamber, 104–5. *See also*
    environmental chamber
Vanguard rocket, 19, 21, 22
Vertical Assembly Building, 222, 223*f*
Vietnam War, 129–30, 147, 158, 195,
    210–11, 212
Vision for Space Exploration, 224–25
Von Braun, Werner
    in Alabama, 12–14
    countering negative space narrative, 83–
        84
    on cover of *Life* magazine, 20–21
    death of, 217
    dialogue between science and
        religion, 83–84
    on Dick Cavett show, 216–17
    on electrical workers strike, 76
    engineering skills of, 234
    entering lunar landing model, 91
    Faget and, 172
    family of, 210

father of, 12
at Fort Bliss, 1, 2
on the future of space program, 129
Gilruth on, 53
on hydrogen fueled rockets, 74
with John F. Kennedy, 82*f*
joining NASA, 25
Kraft and, 53
on landing on the Moon, 48
love-frustration relationship with
    Alabama home, 77
on Mercury program, 87
as national figure, 12, 24
Nazi past of, 1–2, 98, 216–17
at Newcastle Army Base, 1
on non-rocket technology, 50
on orbital spying, 14
organization structure, 49–50, 234
Patriot of the Year award, 24
photograph of, 13*f*
Pohl and, 49, 50
pop-in approach to management, 50
against racism, 77, 97
reassigned to NASA headquarters, 210
retiring from NASA, 210–11
on Saturn V maiden voyage, 145
shuttle program support, 170–71
speaking on new projects, 172
supporting one-ship approach, 68
surrendering to American forces, 1
tempering media on projected Moon
    landing, 180
temper of, 50–51

on "Trip around the Moon," 13–14
unflattering facts on, 83
on welding process, 128
Worlund and, 85–86
writing for general public, 13–14
during WWII, 3–4
Voskhod spacecraft, 83
Voyager project, 171

Wallace, George, 96–97, 151
Wallops Island, 22
Webb, James, 44, 96–97
Weber, Joe, 209
weightlessness in space, 22, 25–26
welding process, clean room environment
    for, 128
Wells, Cynthia, 108, 201, 213
White, Ed, 104, 136
White Sands Proving Grounds, 3*f*
"Why Land on the Moon" (pamphlet), 80
Wolfe, Tom, 54, 66
World War II
    Faget during, 4–6
    German U-boats in Gulf of Mexico, 10
    kamikaze attacks, 4
    von Braun during, 3–4
Worlund, Len, 85–86
Wussler, Robert, 180

Yaw rotation, 69
Young, Ken, 72, 112

Zond spacecraft, 133, 149, 157

Printed in the USA/Agawam, MA
August 15, 2023

814601.037